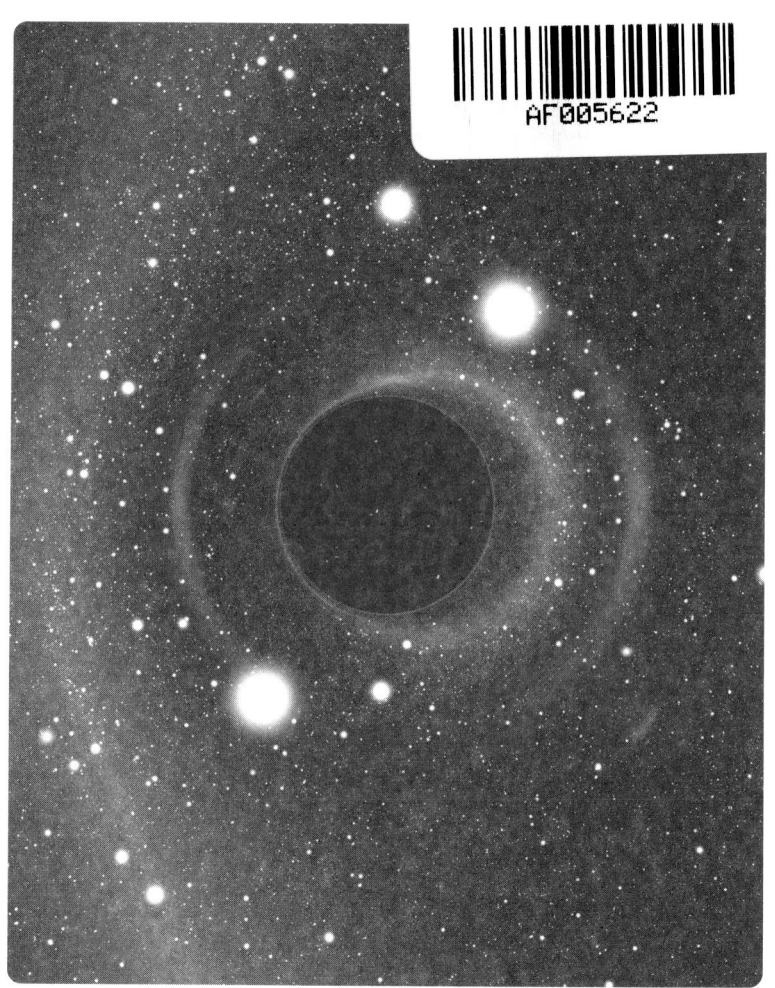

First published 2025
This edition © Wooden Books Ltd 2025

Published by Wooden Books Ltd.
Glastonbury, Somerset.
www.woodenbooks.com

British Library Cataloguing in Publication Data
Linton, O.
Spacetime and Relativity

A CIP catalogue record for this book
may be obtained from the British Library.

ISBN-10: 1-907155-63-5
ISBN-13: 978-1-907155-63-5

All rights reserved.
For permission to reproduce any part of this
timely little book please contact the publishers.

Designed and typeset in Glastonbury, UK.
Printed in India on FSC® certified papers by
Quarterfold Printabilities Pvt. Ltd.

Spacetime & Relativity

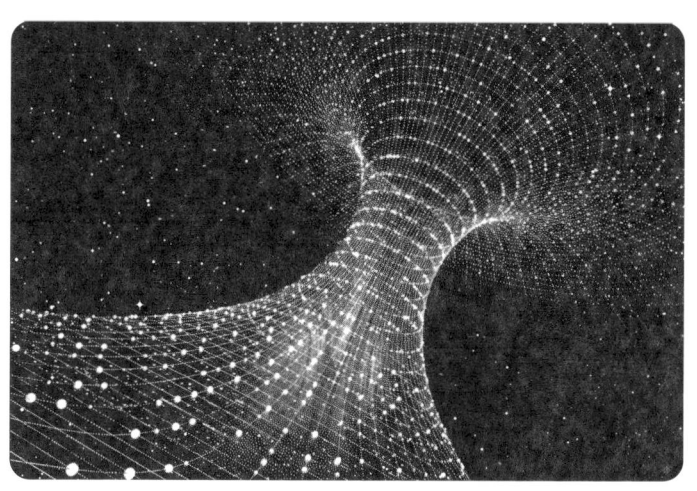

Oliver Linton

One of the most reliable books on the theories of Special and General Relativity is Einstein's own account, written in 1920: *Relativity—The Special and the General Theory*. Another excellent book is *A Journey into Gravity and Spacetime* by Einstein's worthy successor John Archibald Wheeler.

FURTHER READING: Tatsu Takeuchi, *An Illustrated Guide to Relativity*; Martin Gardner, *Relativity Simply Explained*; Lewis Carroll Epstein, *Relativity: Visualized*; Carlo Rovelli, *White Holes*; Sander Bais, *Very Special Relativity: An Illustrated Guide*; N. David Mermin, *It's About Time: Understanding Einstein's Relativity*; Guy Murchie, *Music of the Spheres*.

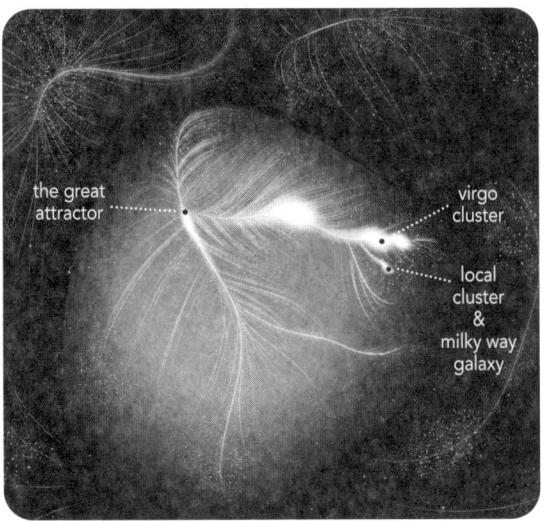

ABOVE: *Our local supercluster Laniakea, showing the direction of motion of its galaxies. An ultra-dense concentration of galaxies, known as the Great Attractor, warps spacetime in a vast gravitational well.*

Contents

Introduction	1
The Speed of Light	2
Coordinate Systems	4
The Galilean Transform	6
Scissored Space	8
The Spacetime Interval	10
The Lorentz Factor	12
Time Slows, Rulers Shorten	14
Past, Present & Future	16
Simultaneity	18
Relativistic Travel	20
The Twin Paradox	22
The Pole & Barn Paradox	24
Bell's Spaceship Paradox	26
Combining Velocities	28
Mass, Momentum & Energy	30
General Relativity	32
Accelerating	34
Curved Space	36
Spacetime	38
Gravitational Time Dilation	40
Geodesics	42
Gravitational Lensing	44
Relativistic Orbits	46
Time Dilation Near a Star	48
The Schwarzschild Radius	50
Black Holes	52
Into a Black Hole	54
Special Effects	56
Appendix: Relativistic Formulæ	58

ABOVE: Light from distant galaxies gravitationally lensed by the curvature of spacetime.

INTRODUCTION

In the final decade of the nineteenth century, physicists were feeling pretty pleased with themselves. Isaac Newton had solved the problems of motion and gravity, and James Clerk Maxwell had sewn up electromagnetism and the nature of light. There seemed to be just three little problems left. The first concerned the colour of a red-hot poker—classical theory predicted that it ought to emit ultraviolet light, not just visible light. (This problem eventually led to quantum theory.) The second was to explain how uranium salts could apparently give off a continuous unlimited stream of radiation—where was the energy coming from? (This problem led to the discovery of the inner workings of the atom.) Beside these two obvious practical deficiencies in classical theory, the third problem seemed to be more of a theoretical puzzle rather than serious weakness of the theory. The trouble concerned the speed of light.

In 1864, Maxwell had unified the study of electricity and magnetism and, in the process, predicted that a system of electric and magnetic fields would propagate itself as an electromagnetic wave through empty space at a speed of about 186,000 miles per second. Since it was known that light travelled at precisely this speed, the inference was obvious—light was an electromagnetic wave capable of travelling through a vacuum. The problem was that nobody could quite believe that any kind of wave could travel through a complete vacuum. Instead, it was proposed that the vacuum was in fact filled with a substance called the 'luminiferous aether', and the race was on to discover the properties of this elusive substance.

THE SPEED OF LIGHT
always the same for any observer

IT WAS GALILEO who first entertained the possibility that light may not travel instantaneously fast. The first evidence of this came from a 1676 study of the orbital periods of one of the moons of Jupiter by the Danish astronomer Ole Rømer. Whenever Jupiter was on the far side of the Solar System from Earth, the moon Io seemed to lag behind by 16 minutes—this being the time it takes for light to cross the diameter of the Earth's orbit. Laboratory experiments by the French physicist Hippolyte Fizeau in the mid-19th century produced reasonably accurate measurements of the speed of light, both in a vacuum and through various transparent substances (*opposite top*).

But did light really travel through some kind of aether? Since, in its annual orbit round the Sun, the Earth must pass through this aether, it followed that the speed of light in a particular direction, as measured on Earth, would change with the planet's orientation. In 1887 two American scientists Albert Michelson and Edward Morley set up a very sensitive experiment to detect this difference in speed. But, to their astonishment, the speed of light seemed to be unaffected by the Earth's motion. The idea that the Earth was the only object in the universe which was stationary with respect to the aether was rejected. But how else could the constancy of the speed of light be explained?

The answer put forward in 1905 by a 26-year-old patent clerk called Albert Einstein was radical. There is no aether. The speed of light through a vacuum is a fundamental constant, the same for all observers—even those in relative motion. Space and time are not separate entities but are intimately woven into a fabric called *spacetime*.

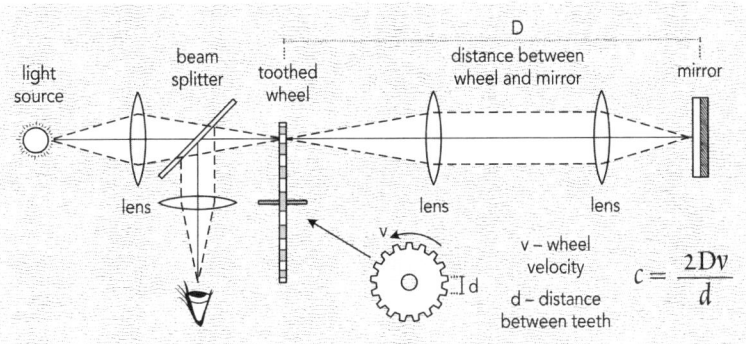

ABOVE: The 1848 Fizeau experiment was the first reasonably accurate measurement of the speed of light, c, in air. Pulses of light were generated by a spinning toothed wheel, which were bounced off a mirror through lenses 8 km apart. The wheel's speed was adjusted until the returning pulses were seen through the next gap, giving $c = 3.13 \times 10^8$ km/s, about 5% too high.

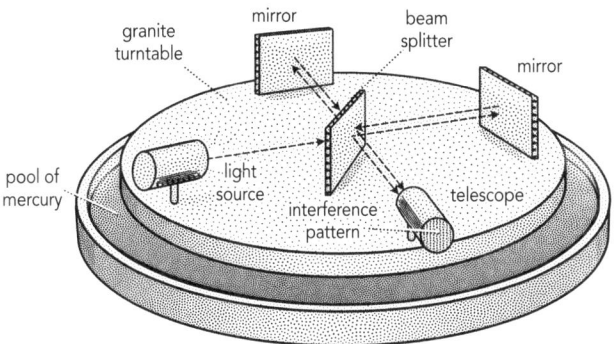

ABOVE: The 1887 Michelson-Morley experiment. If light moved in an 'aether', its speed would be affected by Earth's motion through that aether, boosting or retarding one of the beams, and shifting the interference pattern as the apparatus was turned. The experiment was repeated at different times of the year to see if the Earth's annual motion around the Sun would make a difference, but no shift in the interference pattern was ever observed.

Coordinate Systems
plotting time & space

Maps are representations of where places are in space. They often use a grid reference system whose axes are aligned east-west and north-south; e.g. using the system of latitude and longitude, Westminster Abbey in London is located at 51.5° north, −0.13° east (*below left*).

Imagine the Greek gods have ordained a coordinate system based on the directions 'towards Olympus' and 'across Olympus'. This would still have a rectangular grid but one rotated clockwise by $\frac{1}{10}$ of a turn. Westminster Abbey now has different coordinates, even though it hasn't moved! What is more, the distance between two points, as calculated with Pythagoras' theorem, would remain the same (*below right*).

Now consider a scenario in both time *and* space. Amy flies past Bob, who is standing on a planet (*lower opposite*). From Bob's perspective, or FRAME OF REFERENCE, he is stationary and it is Amy who moves. Their positions can be plotted as *worldlines* on a SPACETIME DIAGRAM. Bob, who does not move through space, is represented by a vertical worldline; Amy, who moves with constant velocity, by a tilted one.

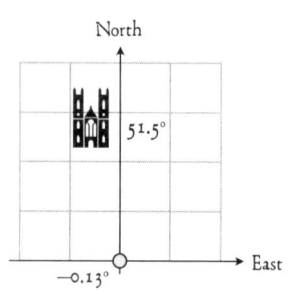 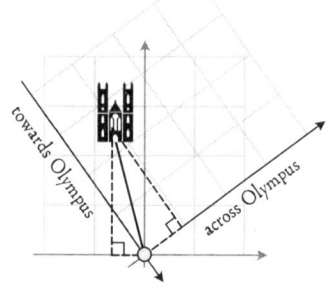

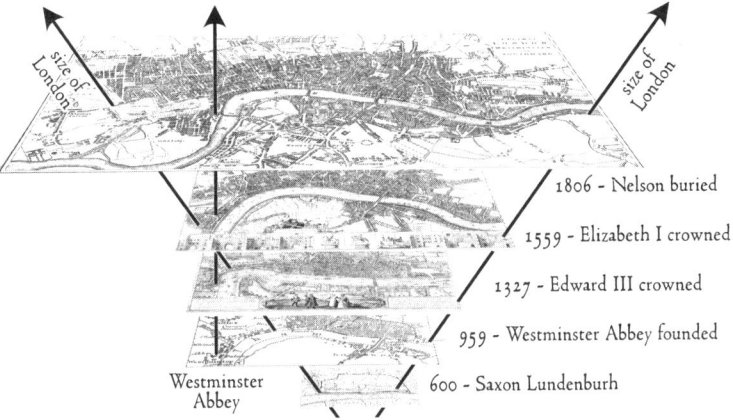

ABOVE: London in spacetime. Westminster Abbey's journey through time is traced by a neat vertical line as it remains stationary in space. The edge of London, moving over time as the city grows, is shown as diagonals, denoting change in both space and time coordinates.

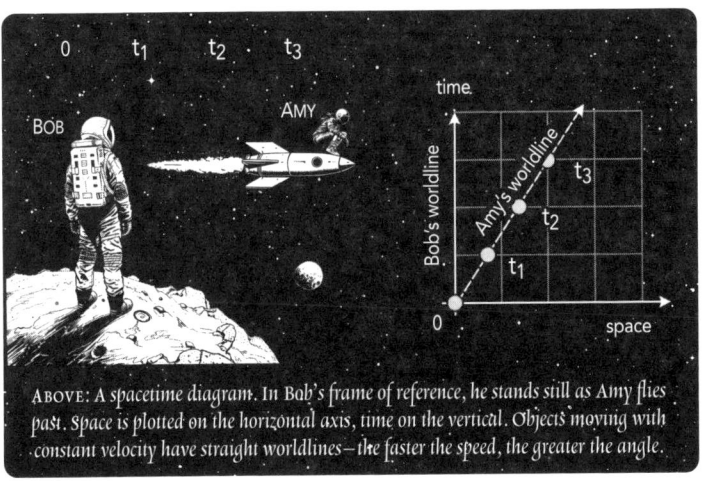

ABOVE: A spacetime diagram. In Bob's frame of reference, he stands still as Amy flies past. Space is plotted on the horizontal axis, time on the vertical. Objects moving with constant velocity have straight worldlines—the faster the speed, the greater the angle.

The Galilean Transform
relative points of view for slow movers

Some time t after Bob saw Amy fly past him, he sees an asteroid smash into the moon she is headed for, a moon which is a distance x from his position on Earth (*below*). According to Bob, the (time, distance) coordinates of this collision are (t, x). Amy sees the impact too. She has been flying at speed v towards the moon, covering a distance vt by the time of the collision, which from her point of view happens at $(t, x-vt)$.

Amy and Bob have two different coordinate systems (*opposite top*), and these can be superimposed in two different ways: Amy's on Bob's, or Bob's on Amy's. Since they can use the same time coordinates, they agree on *when* the smash happened, and so share the same spatial grid lines (horizontal), but they have different experiences of its distance, and so from Bob's perspective Amy's temporal grid lines are tilted to the right. Luckily, however, they can convert between their two coordinates by using simple equations (*lower opposite*).

These equations are known as the GALILEAN TRANSFORMATION, because Galileo Galilei [1564–1642] was the first person to realize that there is no such thing as absolute motion: only *relative* motion.

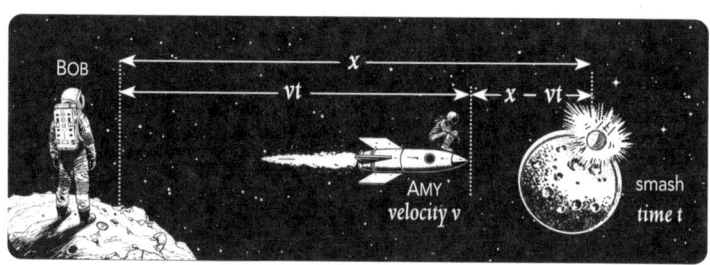

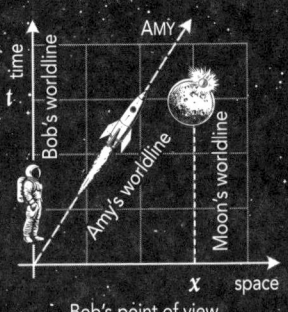

Bob's point of view

ABOVE: In Bob's frame of reference, he and the moon are stationary and Amy is moving. Bob sees the smash at (t, x).

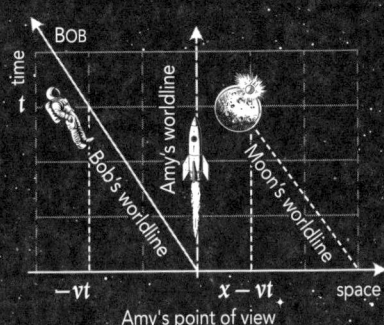

Amy's point of view

ABOVE: In Amy's frame she is stationary, while Bob and the moon are both moving. For Amy, the smash occurs at $(t, x - vt)$.

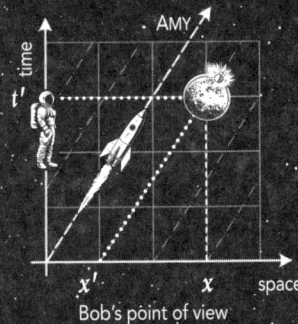

Bob's point of view

BOB converts to AMY's frame of reference:
$t' = t \quad x' = x - vt$

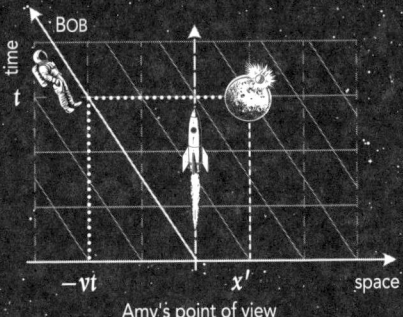

Amy's point of view

AMY converts to BOB's frame of reference:
$t = t' \quad x = x' + vt'$

ABOVE: Spacetime diagrams of Amy and Bob's experiences of the asteroid collision. The only difference between Bob and Amy's views is that the time axis gets tilted over one way or the other. At low velocities, t and t' are equal, so both can use the same spatial axis.

SCISSORED SPACE
Minkowski diagrams

The Galilean transformation works if time passes uniformly for *all* observers. However, this is not the case as we speed up. Since the speed of light, c, is constant for all observers, in his SPECIAL THEORY OF RELATIVITY Einstein suggested that not only does the time axis tilt, but the space axis tilts as well, symmetrically about a 45° line representing the speed of light (like a pair of scissors, *see opposite*). In this way, both the temporal and spatial coordinates of any event which lie on the 'speed of light' line are multiplied by the same factor and so every observer then measures the same value for the speed of light, however fast they travel. Mathematically, this means adopting the following transformation:

$$t' = \gamma\left(t - vx/c^2\right) \qquad x' = \gamma\left(x - vt\right)$$

where γ (the Greek letter *gamma*) is the *Lorentz factor* (*see page 12*), and c is the speed of light. Although Amy and Bob use different coordinate systems, they must always agree on the speed of light. If Bob finds that light travels a distance x in time t (where $x = ct$), then Amy finds it travels a distance $x' = \gamma\left(ct - vt\right)$ in time $t' = \gamma\left(t - vct/c^2\right)$. She will then also calculate the speed of light to be:

$$\frac{x'}{t'} = \frac{\gamma(ct - vt)}{\gamma(t - vct/c^2)} = \frac{c - v}{1 - v/c} = c$$

provided that both Amy and Bob are in INERTIAL REFERENCE FRAMES, i.e. moving at constant speed in straight lines relative to each other.

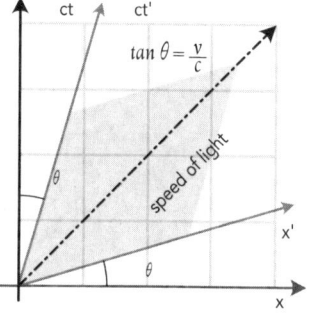

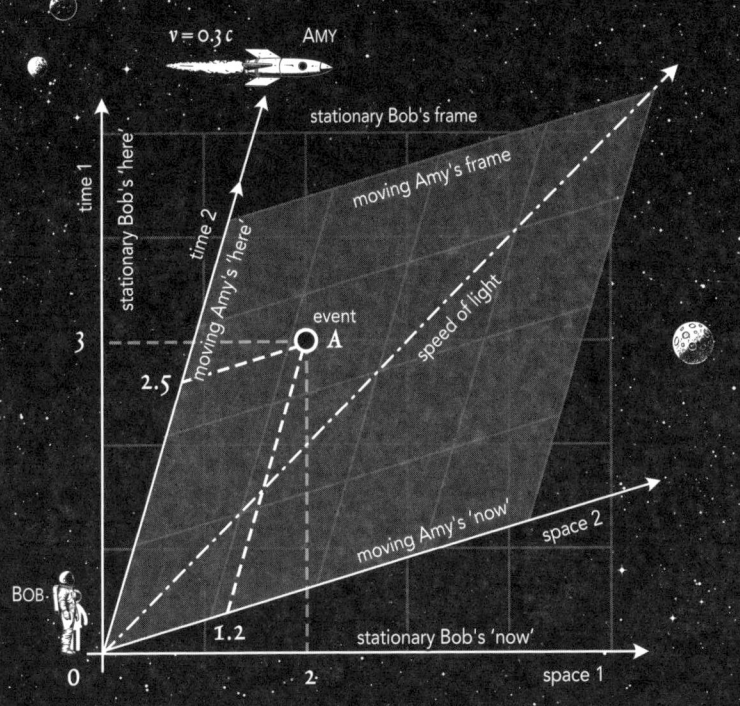

ABOVE & OPPOSITE: Minkowski diagrams are scaled so that the worldline representing the speed of light c is inclined at 45°. The vertical scale is set to ct, the time light takes to travel one unit of distance x (horizontal scale), using units such as years and light-years where c = 1. The example above shows the coordinate system of a spaceship travelling at 30% of the speed of light. The Time 2 axis joins all events which occur at the same place (e.g. 'here', inside the spaceship), while the Space 2 axis joins all events which occur at the same time (e.g. the spaceship's 'now'). Notice how the marked event A occurs at coordinates (3 years, 2 light-years) in the stationary frame of reference, but at (2.5 years, 1.2 light-years) in the tilted moving frame.

THE SPACETIME INTERVAL
the proper time

Amy and Bob always measure the same value for the speed of light, no matter how fast they are moving (the first principle of special relativity). The second principle of special relativity states that Amy's point of view is just as valid as Bob's, so the transform which Bob uses to turn his (t, x) frame coordinates into Amy's (t', x') frame coordinates must work in the reverse direction:

Transform Bob's to Amy's: $\quad t' = \gamma(t - vx/c^2) \quad x' = \gamma(x - vt)$

Transform Amy's to Bob's: $\quad t = \gamma(t' + vx'/c^2) \quad x = \gamma(x' + vt')$

Notice how Bob's (t, x) coordinates are expressed in terms of Amy's (t' and x') and how the sign of v is changed. Combining these equations with a little bit of algebra now reveals the SPACETIME INTERVAL, s:

$$s = \sqrt{t^2 - x^2/c^2} \; (recorded\ by\ Bob) = \sqrt{t'^2 - x'^2/c^2} \; (recorded\ by\ Amy)$$

The spacetime interval measures the 'distance' in units of time of an event at coordinates (t, x) from the origin, and is the same for *all* observers, if they are in inertial (non-accelerating) frames of reference.

Imagine Amy has reached a point (t, x) in Bob's frame. She regards herself as being at the point $(t', 0)$ in her own coordinate system because, according to her, she has not moved. She therefore calculates $s = \sqrt{t'^2 - 0} = t'$. This is Amy's PROPER TIME—her journey time as recorded by her ship's clocks. Since the spacetime interval between two events is the same for all observers, Bob can calculate Amy's proper time to be $t' = \sqrt{t^2 - x^2/c^2}$, which simplifies to $\sqrt{t^2 - x^2}$ if t is in units of years and x is in units of light-years (*see opposite*).

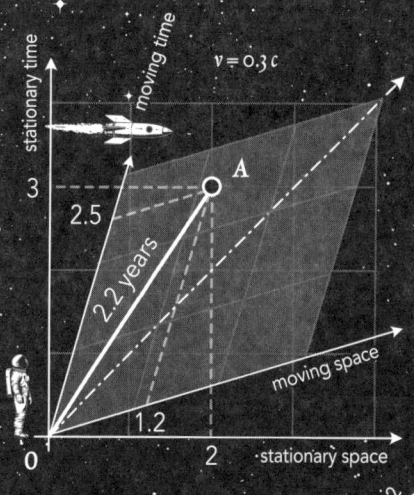

LEFT: Proper time: In the 'stationary' frame, the coordinates of event A (in years and light-years) are (3, 2). The proper time between event A and the origin O is: $\sqrt{3^2 - 2^2} \approx 2.2$ years. This is the actual journey time as measured by clocks on board. In the 'moving' frame, the coordinates are (2.5, 1.2). And once again the proper time is: $\sqrt{2.5^2 - 1.2^2} \approx 2.2$ years.

RIGHT: The angle of the worldline representing an object's trajectory through spacetime changes as its speed increases, tilting it towards the 45° speed of light line.

A key property of the spacetime interval is that it is invariant for all inertial observers, remaining constant even if they are in motion, and measure different distances and times.

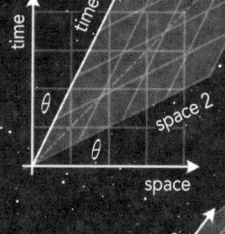

$v = 0.5c$
slope = 2:1
$\theta = 26.6°$

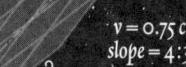

$v = 0.75c$
slope = 4:3
$\theta = 36.8°$
where $\tan \theta = \dfrac{v}{c}$

The Lorentz Factor
the faster you go

The position of an event in spacetime is represented by a set of four coordinates: t, x, y, and z. The full expression for the spacetime interval s between an event and the origin of the coordinate system is:

$$s = \sqrt{t^2 - (x^2 + y^2 + z^2)/c^2}, \text{ where } c \text{ is the speed of light.}$$

However, if motion is assumed to be confined to the x-axis, this allows y and z to be set to zero, resulting in a simpler: $s = \sqrt{t^2 - x^2/c^2}$.

Our four equations (*page 10*) can be used to calculate an expression for the Lorentz factor γ. Substituting for x' and t' into $x = \gamma(x' + vt')$ gives:

$$x = \gamma(\gamma(x - vt) + v\gamma(t - xv/c^2))$$
$$= \gamma^2(x - vt + vt - xv^2/c^2) \quad = \gamma^2 x(1 - v^2/c^2)$$

$$\text{therefore: } \gamma = \frac{1}{\sqrt{1 - v^2/c^2}}$$

Thus γ, the constant of proportionality which makes the transform symmetrical, is dependent on the relative velocity v in an interesting way. When the velocity v is much smaller than the speed of light c we can ignore the term v^2/c^2 which makes $\gamma \approx 1$, but as v increases, this term shrinks, letting γ rise to infinity as v tends to c (*opposite top right*).

Derived by Woldemar Voigt in 1887 and Hendrik Lorentz in 1904, the Lorentz factor γ and its various transformations demonstrate how the constancy of the speed of light results in moving clocks running slow and moving rulers getting shorter. Equations and graphs for the fundamental transformations are shown (*lower opposite*). Some illustrative examples appear on the next page.

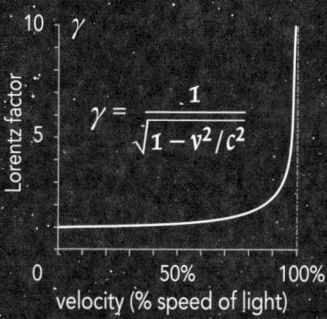

ABOVE: As an object moves faster and faster the term v^2/c^2 gets closer and closer to 1 and the Lorentz factor γ increases, becoming infinite when the object reaches the speed of light.

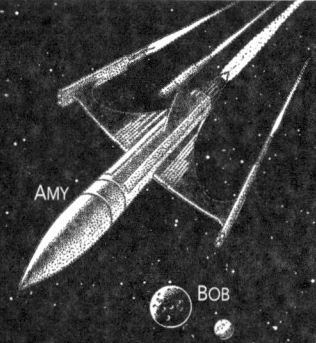

ABOVE: A remarkable consequence of Special Relativity is that, from Bob's point of view, Amy's spaceship is shortened and her clocks appear to go slow (see below and next page).

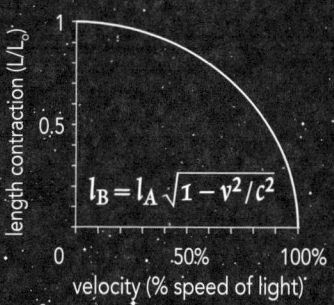

ABOVE: Length contraction: The faster Amy moves, the shorter her spaceship appears to Bob. If the length of the spaceship is l_A as measured by Amy on board, then Bob will measure its length as it flies by to be l_A/γ.

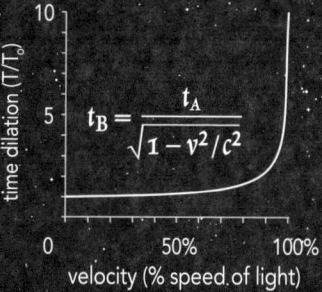

ABOVE: Time dilation: The faster Amy moves, the slower her clocks appear to run as seen by Bob. If it takes t_A minutes for Amy to boil an egg on board, then Bob will reckon that it has taken her $t_A \gamma$.

TIME SLOWS, RULERS SHORTEN
the Lorentz factor in action

Amy sets up a simple clock which bounces a photon of light between two parallel mirrors set up across her spaceship at distance d_A apart. This ticks at a regular interval, every time it hits a mirror, observed by Amy as $t_A = d_A/c$, where c is the speed of light. Amy and Bob agree about the distance d_A. However, as Amy flies by at a velocity v, Bob sees the photon moving at light speed c along a longer diagonal path, length d_B (*opposite top*). By Pythagoras' theorem:

$$d_B^2 = (ct_B)^2 = (vt_B)^2 + d_A^2$$

where t_B is the time between ticks according to Bob. Light can't speed up or slow down, so rearranging this equation, we obtain:

$$t_B = d_A/\sqrt{c^2 - v^2} = ct_A/\sqrt{c^2 - v^2} = t_A/\sqrt{1 - v^2/c^2} = t_A\gamma.$$

and see that time slows down—Bob sees Amy's clock running slow!

Amy now rotates her clock so it is parallel to the motion of the ship (*lower, opposite*). Although she still sees regular ticking, Bob observes uneven ticks as she flies by, for when the photon is travelling in the same direction as the spaceship, the tick is longer, at $d_B/(c-v)$, but on bouncing back, the tick is shorter, at $d_B/(c+v)$. Bob adds the two times:

$$2t_B = d_B/(c+v) + d_B/(c-v) = 2d_B c/(c^2 - v^2) = 2d_A\gamma^2/c.$$

Amy and Bob both see the photon travel different distances but return to the first mirror at the same time. So $2t_A\gamma = 2d_A\gamma/c = 2d_B\gamma^2/c$, and $d_B = d_A/\gamma$. Light can't speed up or slow down, so Bob sees Amy's spaceship and everything in it shortened in the direction of motion.

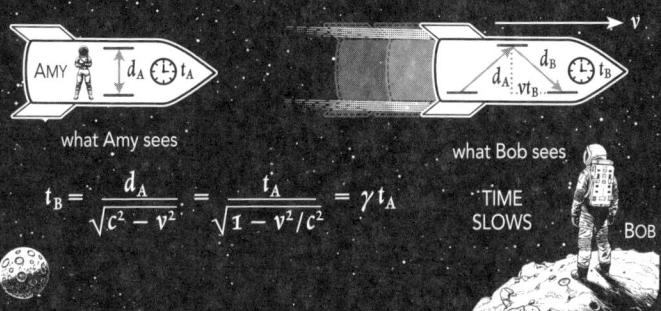

$$t_B = \frac{d_A}{\sqrt{c^2 - v^2}} = \frac{t_A}{\sqrt{1 - v^2/c^2}} = \gamma\, t_A$$

TIME SLOWS

ABOVE: Amy has a light clock which bounces a photon vertically between two mirrors, distance d_A apart, at d_A/c ticks per second. As she flies past Bob, at velocity v, he sees the photon travel a longer zig-zag path, distance d_B, in the same time. Light cannot speed up, so time must slow down. Bob sees Amy's clock with a longer tick, by the Lorentz factor γ.

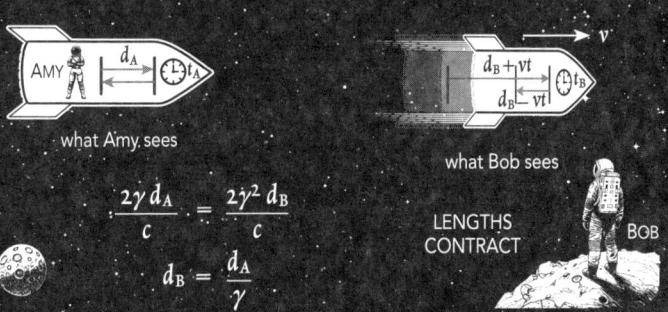

$$\frac{2\gamma\, d_A}{c} = \frac{2\gamma^2\, d_B}{c}$$

$$d_B = \frac{d_A}{\gamma}$$

LENGTHS CONTRACT

ABOVE: Amy rotates her clock, so the photon now travels along the direction of motion. From Bob's point of view, when the photon is moving to the right it is trying to catch up with the spaceship which is moving away from it. The relative velocity between the photon and the spaceship is $c - v$. On the other hand, when the photon is travelling back to the left, the relative velocity is $c + v$. Bob sees Amy's ship shortened by the Lorentz factor γ.

Past, Present & Future
this wide and universal theatre

In a spacetime diagram, the horizontal space axis joins all events which occur *at the same time* in a frame of reference—let's call it the 'now' line. As we have seen, an object moving in the positive x direction causes this 'now' line to tilt upward, while moving in the negative direction tilts it down, limited only by the speed of light's 45° lines. Spacetime can therefore be divided into three distinct zones, bounded by LIGHT CONES (*opposite top*). These regions are:

LIGHTLIKE or null; $s = 0$ (where s is the spacetime interval). This is the light cone's boundary, the speed of light. Events are linked only by massless objects like photons that move at the speed of light.

TIMELIKE; $s^2 > 0$: Particles with mass (i.e. moving slower than light) can attend all events. Events are *causally connected*. Observers agree about order and can influence events. The upper cone contains the *absolute future*—all events which we could potentially reach; the lower cone is the *absolute past* with all the events *from* which we could have travelled (at below light speed).

SPACELIKE; $s^2 < 0$: Events are so far apart that no signal, even at the speed of light, can ever travel from one event to the other. Events are *causally disconnected* and observers may disagree about the order of events. In between the cones is a grey area which is neither future nor past—the *potential present*. We cannot travel to events in its region above the 'now' line—the *inaccessible future*—because to get there we would have to travel faster than light. Likewise, events below the 'now' line—the *inaccessible past*—cannot influence us (*lower opposite right*).

Thus, spacetime diagrams show who can see what, and when.

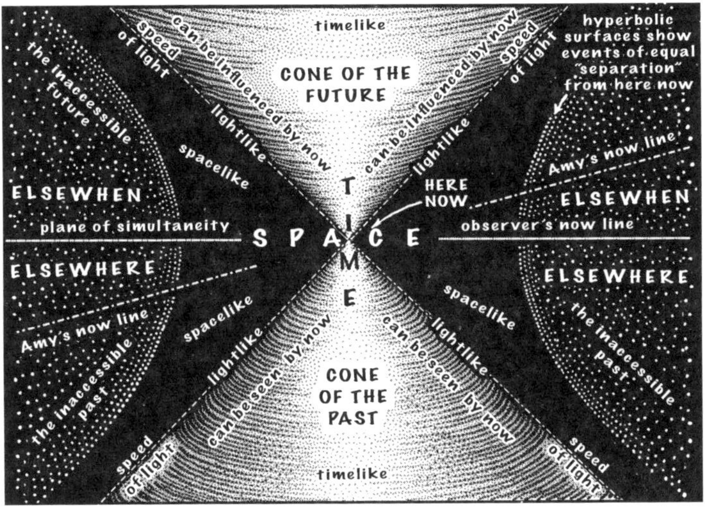

ABOVE: Light cone diagrams add an extra spacial dimension to a spacetime diagram to show 3D space as a 2D horizontal plane, with time as the vertical axis. The cones meet at the observer's 'now'—the plane of simultaneity. Within are 'timelike' regions which are causally connected. Outside the cones, 'spacelike' regions are causally disconnected. After Guy Murchie.

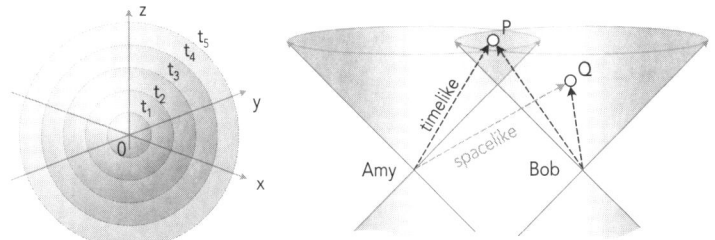

ABOVE: Light cones are 2D representations of spherical shells of light spreading through 3+1D spacetime from the observer's present 'now' moment located at the origin.

ABOVE: Event P has timelike separation and can be experienced by both Amy and Bob. Event Q has spacelike separation with respect to Amy and can only be experienced by Bob.

SIMULTANEITY
or the lack of it

In the framework of special relativity, the concept of 'now' or the present is not a universal constant but is dependent on the observer's frame of reference, in particular their state of motion.

The RELATIVITY OF SIMULTANEITY can be illustrated by imagining two stars exploding just as Amy and Bob are equidistant from both. Bob, on an asteroid, is stationary relative to the stars and sees the two flashes simultaneously. Amy, however, is in her moving spaceship, so her 'now' line is tilted upwards, and to her, the explosion in front has already happened, while the rear one is yet to take place (*opposite top*).

It is important to distinguish what Amy actually *sees* from what she *deduces* has happened. Since Amy and Bob are in the same place at the same time, both see the two flashes simultaneously. But Amy, taking into account the finite speed of light, deduces that—in her frame of reference—the front flash happened before the rear one because she is moving towards it. Conversely, a third observer moving rapidly from right to left at the same instant would deduce that the rear flash was first. However, there is no inconsistency, as the two events are separated by a *spacelike* interval and there is no way that P and Q can be causally connected.

If, on the other hand, two events P and Q are causally connected, they must have a *timelike* separation and in this case, if P causes Q, then all observers will see P happening before Q.

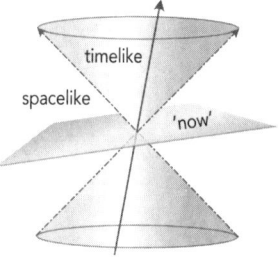

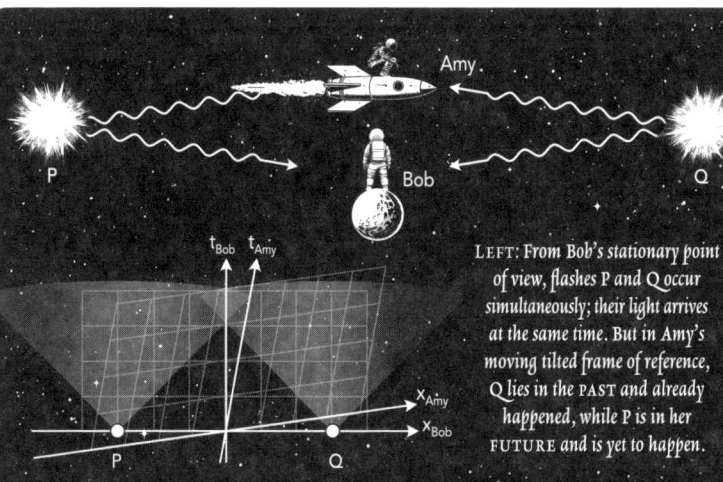

LEFT: From Bob's stationary point of view, flashes P and Q occur simultaneously; their light arrives at the same time. But in Amy's moving tilted frame of reference, Q lies in the PAST and already happened, while P is in her FUTURE and is yet to happen.

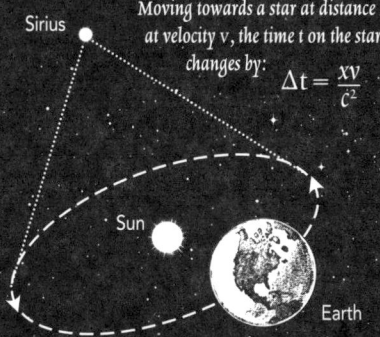

RIGHT: Sirius is 8600 light-years away from Earth, which orbits the Sun at 1/10,000th the speed of light. As Earth moves towards Sirius, in Earth's frame of reference the time on the star appears to move forward by about 314 days. Six months later, as Earth moves away, time appears to fall back by 314 days. In reality, nothing changes on Sirius; this effect is simply due to the shift in Earth's frame of reference, which tilts the 'now' line.

Moving towards a star at distance x at velocity v, the time t on the star changes by:
$$\Delta t = \frac{xv}{c^2}$$

FACING PAGE: Relative motion affects the perception of time by tilting the 'now' plane of a light cone in the direction of travel, changing which events are considered simultaneous by different observers. The faster the speed, the greater the shift in perceived time. For distant objects, even small differences in speed can lead to significant changes in the perception of when certain events occur.

Relativistic Travel
the long & short of it

Humans have long dreamt of journeying to the stars. The next star is Proxima Centauri, 4 light-years away. However, even if we used the motion of the Earth to 'sling-shot' our way there at 1/10,000th of the speed of light, it would still take 40,000 years to get there!

What if Amy was able to travel at a sizeable fraction of the speed of light, say 80%? Then she should only take $4 \div 0.8 = 5$ years to get there. According to people at the origin on Earth, her arrival at the star will have coordinates $(t, x) = (5, 4)$. But there's a twist, the proper time (*see page 10*) between the origin and $(5, 4)$ is $\sqrt{5^2 - 4^2} = 3$ years, so this will be the time that passes for Amy on the spacecraft.

But that is crazy—how can you get to a star 4 light-years away in only 3 years? Nothing can travel faster than light! The answer is that, to Amy on the spaceship, the distance between Earth and the star has been contracted by a factor of γ which, in this instance, is $\frac{1}{\sqrt{1 - 0.8^2}} = 5/3$. That makes the distance travelled equal to $4 \times 3/5 = 2.4$ light-years which, at a speed of 0.8 light-years per years takes exactly 3 years (*below*).

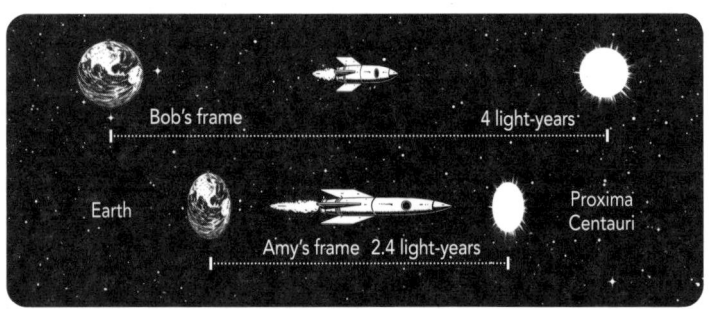

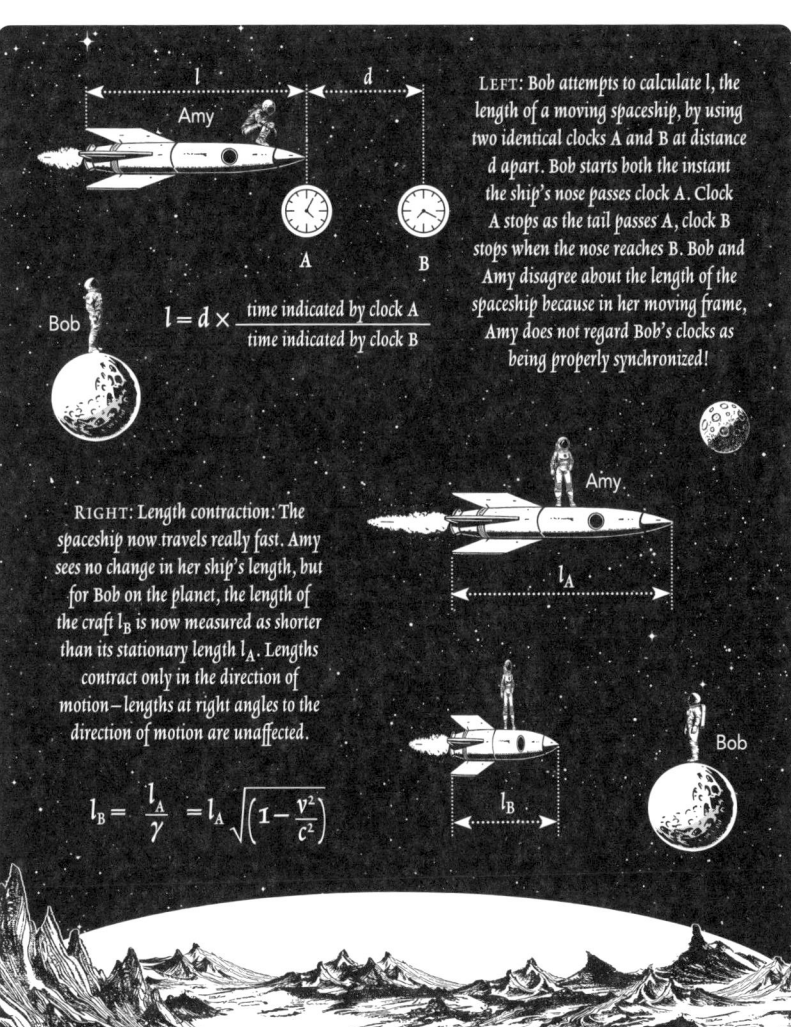

LEFT: Bob attempts to calculate l, the length of a moving spaceship, by using two identical clocks A and B at distance d apart. Bob starts both the instant the ship's nose passes clock A. Clock A stops as the tail passes A, clock B stops when the nose reaches B. Bob and Amy disagree about the length of the spaceship because in her moving frame, Amy does not regard Bob's clocks as being properly synchronized!

$$l = d \times \frac{\text{time indicated by clock A}}{\text{time indicated by clock B}}$$

RIGHT: Length contraction: The spaceship now travels really fast. Amy sees no change in her ship's length, but for Bob on the planet, the length of the craft l_B is now measured as shorter than its stationary length l_A. Lengths contract only in the direction of motion—lengths at right angles to the direction of motion are unaffected.

$$l_B = \frac{l_A}{\gamma} = l_A \sqrt{\left(1 - \frac{v^2}{c^2}\right)}$$

THE TWIN PARADOX
there & back again

The fact that Amy can get to a star 4 light-years away in only 3 years is strange enough—but what is even stranger is that if she turns round and heads back home (another 3 year journey), she will find that Bob, her twin, has aged by 10 years and is now 4 years older than her! Bob saw Amy travel a total of 8 light years at 80% of light speed, which took 10 years. But only 6 years have passed for her. So Bob reckons that time on Amy's spaceship slowed by a factor of 5/3—the same γ factor that made her journey to the star shorter than expected (*page 20*).

The TWIN PARADOX arises from the fact that relativistic effects are symmetrical. Amy might argue that, from her perspective, it is Earth that has flown away and returned and so it is Bob who should be younger. The paradox is resolved by understanding that an asymmetry arises in the turnaround at the star, since only the traveller changes their coordinate system and experiences acceleration (*opposite*).

Time dilation has been experimentally verified. In 1971, Joseph C. Hafele and Richard E. Keating took four atomic clocks aboard an airliner. They flew twice around the world, first eastward, then westward, and then compared the clocks against others that had remained on the ground (*right*). When reunited, the clocks showed a time difference of a few pico-seconds, entirely consistent with the predictions of relativity.

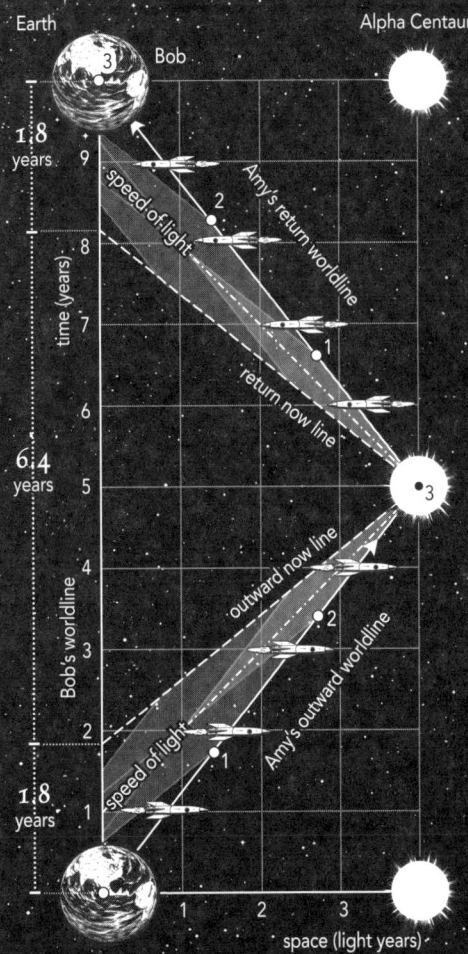

LEFT: **The twin paradox:** Both twins see their sibling's clock run slow. However, as Amy changes direction, the relativity of simultaneity causes a shift in her perception of Bob's clock. From her new coordinate system, Bob's clock appears to advance as a result of her changing between two different inertial frames of reference.

As both are in relative motion, each twin's clocks run slow by a factor of γ. In this case, $\gamma = 5/3$, and so in the 10 years that Bob awaits Amy's return, he calculates that her clock will only have advanced 6 years ($10 \times 3/5$) and indeed finds that Amy is now 4 years younger than him.

For Amy, in the 3 years it takes her to reach Alpha Centauri, she deduces that Bob's clocks will have advanced by 1.8 years ($3 \times 3/5$). Then, as she turns around, she transforms between inertial frames, which means that time on Earth appears to sweep forward by 6.4 years ($2 \times xv = 2 \times 4 \times 0.8$). Finally, she adds in another 1.8 years for Bob's time during her return journey. She therefore correctly expects Bob to have aged 10 years on her arrival home.

THE POLE & BARN PARADOX
when one door closes

Here's a puzzle—a 4.5 m barn has doors at each end. Amy runs through with a 5 m long pole. Can the pole fit into the barn with both doors closed? Simple, you might think! If Amy runs at 60% of the speed of light, length contraction will shrink the pole to only 4 m long, so both doors can be shut for the brief period it is fully inside. That is, from the point of view of Bob standing outside the barn.

But what about Amy's point of view? To her, the barn is racing towards her at 60% of light speed and so it is the barn, not the pole, which is length contracted. In fact, Amy sees the barn as only 3.6 m long and there is no way a 5 m pole will fit (*opposite top*).

In Bob's frame, both doors can indeed be shut for the time the pole is inside. In Amy's frame, although the barn is too short, due to the relativity of simultaneity the doors are not closed concurrently as they are for Bob. To Amy, the left door closes *after* the right, and for her, there is no single moment when the whole pole is inside.

This difference in simultaneity resolves the paradox. From each observers' point of view the object really is shortened in the direction of motion (*below*), just as Amy's clocks really do run slow relative to those of her stay-at-home twin (*page 22*). However, both disagree about the order of events, because their 'now' lines do not coincide.

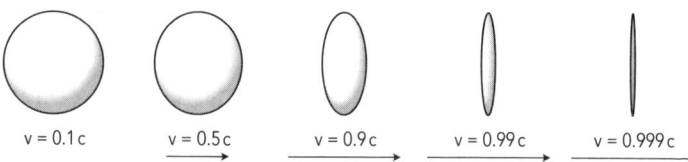

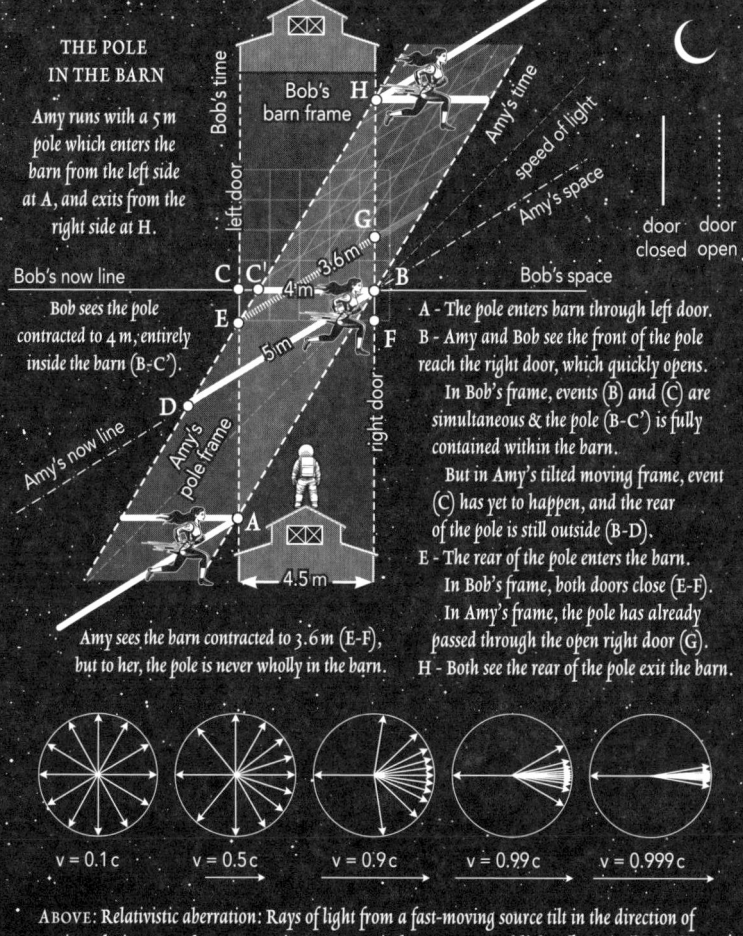

BELL'S SPACESHIP PARADOX
pulling the rope apart

Unlike time dilation, which has been verified in countless experiments involving particles moving at high speeds in an accelerator, it is trickier to directly verify length contraction. On the other hand, since length contraction is a necessary consequence of the constancy of the speed of light, any experiment such as the Michelson-Morley (*page 3*) which confirms the latter also confirms the former.

Imagine two stationary spaceships 100m apart joined by a 100m rope, both equidistant from a red **STOP** light (event O, *opposite*). The light turns green, signalling **GO**, and the moment they see this (events P and Q) both ships fire their warp drives, accelerating instantly to the same very high speed. What happens to the rope? Does it contract and break? Or do the two ships somehow get closer together as they speed up?

For a stationary observer at O, the answer is both ships accelerate equally and the distance between them stays the same. However, the ships and the rope undergo length contraction, and the rope breaks.

For other observers, the answer may be different. Ship C happens to be travelling at the same final speed as A and B, in the same direction, and passes by at the exact instant they start to move (U). Due to C's motion, his line of simultaneity ('now line') is tilted. From C's perspective, event Q (when ship B starts moving) is simultaneous to event S, and so *before* event P (when ship A starts moving) which, for C, is concurrent with event T. Hence, the pilot of ship C also sees the rope break—but for a different reason!

100m

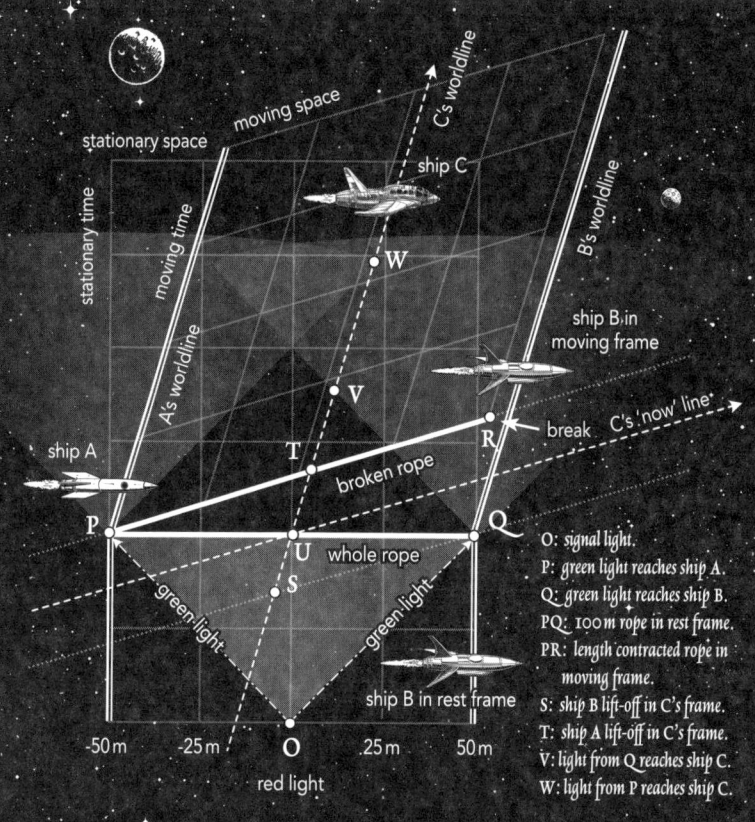

ABOVE: Bell's spaceship paradox: In the stationary frame of the signal at O, the GO light reaches ships A and B simultaneously (events P & Q). Both ships accelerate equally, maintaining a constant distance, but the rope is length contracted and breaks. In ship C's moving frame, at the moment C crosses the line PQ at U, event Q is in the past (S) while event P is yet to happen (T). Ship C's pilot concludes the rope breaks because ship B took off first, and later sees the light from Q at V and P at W.

COMBINING VELOCITIES
observing the speed limit

Why is it impossible to travel faster than light? Surely a missile fired at 80% of light speed from a spaceship which is travelling at 50% of light speed will achieve an overall velocity of 130% light speed?

It turns out that speeds do not add up like this. Suppose Amy, travelling past Bob at a velocity v, fires a gun from the back of her spaceship towards a target at the front, a distance d away (*below*). The gun's muzzle velocity is u. In her frame of reference, the bullet hits the target at the (time, space) coordinates $(d/u, d)$. Using the Lorentz equations (*page 10*), these coordinates in Bob's frame (t, x) are:

$$t = \gamma \left(d/u + vd/c^2 \right) \qquad x = \gamma \left(d + vd/u \right)$$

The speed w of the bullet as seen by Bob is equal to x/t, so:

$$w = \frac{\gamma \left(d + vd/u \right)}{\gamma \left(d/u + vd/c^2 \right)} = \frac{u + v}{1 + uv/c^2}$$

Since both u and v must be less than c, it can easily be seen that the combined speed can never be greater than that of light.

Further equations of relativistic motion are shown opposite.

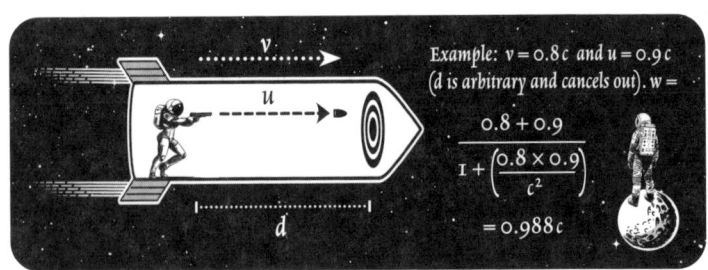

Example: $v = 0.8c$ and $u = 0.9c$
(d is arbitrary and cancels out). $w =$

$$\frac{0.8 + 0.9}{1 + \left(\dfrac{0.8 \times 0.9}{c^2} \right)}$$

$$= 0.988c$$

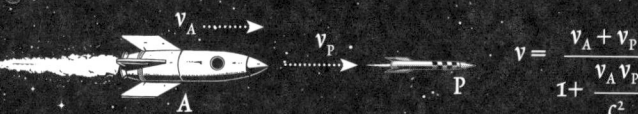

ABOVE: Relativistic velocity addition: If a spaceship travelling at 50% of the speed of light fires a projectile with a speed of 80% of the speed of light, then:
$V_A = 0.5c$ and $V_P = 0.8c$, so the combined speed is $(0.5 + 0.8)/(1 + 0.5 \times 0.8) = 0.93c$

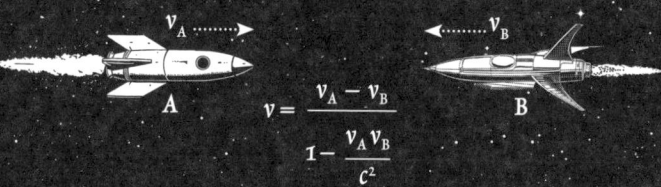

ABOVE: Relativistic approach velocity: If two spaceships travel towards each other at 50% of the speed of light, their combined speed of approach as seen by the occupants will be:
$(0.5 + 0.5)/(1 + 0.5 \times 0.5) = 0.8c$

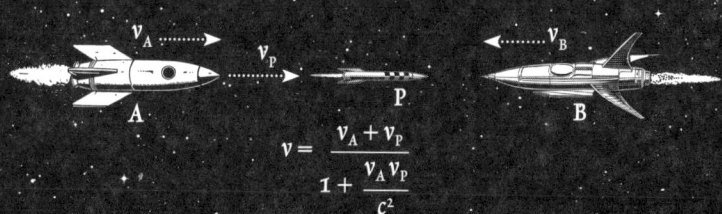

ABOVE: Combined velocities: Since we know that from Bob's point of view the projectile is travelling at $0.93c$, we can just use the top formula to add $0.93c$ and $0.5c$.
This gives us $(0.93 + 0.5)/(1 + 0.93 \times 0.5) = 0.98c$

MASS, MOMENTUM & ENERGY
mc squared

There is another reason why nothing can travel faster than light, and it has to do with mass and energy. When Einstein tried to incorporate the conservation laws of momentum and energy into his theory, it appeared that the momentum of a mass m moving at a velocity v was not mv but γmv (where $\gamma = 1/\sqrt{1 - v^2/c^2}$, see page 12). On the face of it, it appears that the mass of the object increases as it goes faster. But why should motion affect mass?

Einstein's extraordinary insight was that energy had mass! In fact, he determined that the mass of E joules of energy is E/c^2 where c is the speed of light. In other words, the fact that the momentum of a mass m travelling with a velocity v is greater than mv is not really because the mass of the object has increased—rather it is because the extra kinetic energy of the moving object also has mass!

Newton's equation for the kinetic energy (KE) of an object mass m, moving at velocity v was $KE = \tfrac{1}{2}mv^2$. Einstein modified this to:

$Total\ mass = m + KE/c^2$

$Total\ momentum = (m + KE/c^2)v = \gamma mv$

$m + KE/c^2 = \gamma m$

$KE = mc^2(\gamma - 1)$

An odd consequence of this mass-energy equivalence is that a hot cup of tea weighs more than a cold cup, a new battery weighs more than a flat one, and a wound-up watch weighs more than one that has run down!

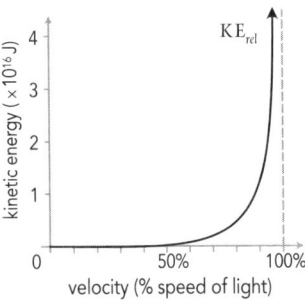

Energy of an object, mass m, in a stationary frame (Rest Energy):

$E_R = mc^2$

Total energy of an object, mass m, moving at velocity v (Rest + Kinetic energy):

$E_T = \gamma mc^2$

ABOVE: Mass has energy: Exploding fireworks convert a tiny fraction of their mass into energy. Atomic bombs are a little more efficient and convert slightly more – about 0.01%.

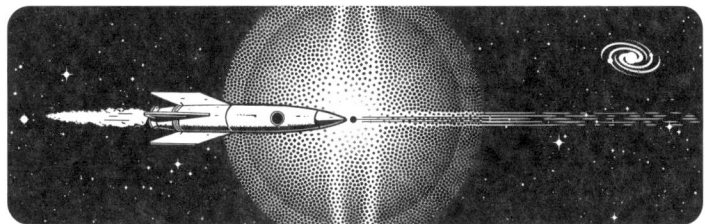

ABOVE: Travelling at relativistic speeds is dangerous – if a spaceship hits a pea at 0.9c, it will yield the same amount of energy as a 10 kiloton nuclear bomb. Particles like photons and EM waves are less hazardous; they are massless and travel at light speed, and therefore do not experience time passing – from a photon's point of view, all distance is covered in an instant.

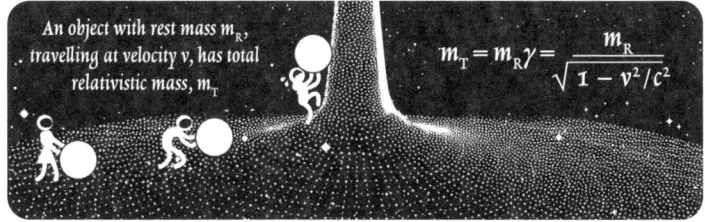

An object with rest mass m_R, travelling at velocity v, has total relativistic mass, m_T.

$$m_T = m_R \gamma = \frac{m_R}{\sqrt{1 - v^2/c^2}}$$

OPPOSITE & ABOVE: Like pushing a rock up an ever steeper slope, the kinetic energy of an massive object mass rises infinitely as it nears light speed, needing increasing amounts of energy to accelerate. External observers will see the object's total relativistic mass (m_T) rise, while its rest mass (m_R) – which is unaffected by motion or frame of reference – is unchanged.

GENERAL RELATIVITY
the equivalence principle

Einstein based his SPECIAL THEORY OF RELATIVITY (*pages 8-31*) on two principles: that the laws of physics are the same for everyone, even if in relative motion, and that the speed of light is constant. The trouble was, it only seemed to apply when moving in straight lines at constant speed. If you accelerated or turned a corner, nothing made sense.

Then he had a brainwave! He realized that an observer in an enclosed spaceship would not be able to tell whether the force which made things fall to the back of the cabin was caused by the ship's acceleration, or whether it was because the ship was standing on a planet with gravity. In other words: it is impossible to distinguish between accelerated motion and a local gravitational field (*opposite*).

The consequences which follow from this *equivalence principle* are far reaching and led Einstein to his GENERAL THEORY OF RELATIVITY: 1: gravity causes light to bend, 2: gravity causes clocks to go slow, and 3: gravity causes spacetime to curve. This last consequence revolutionized our concept of gravity—instead of being an attractive force propagating instantaneously through space, it becomes a feature of space itself.

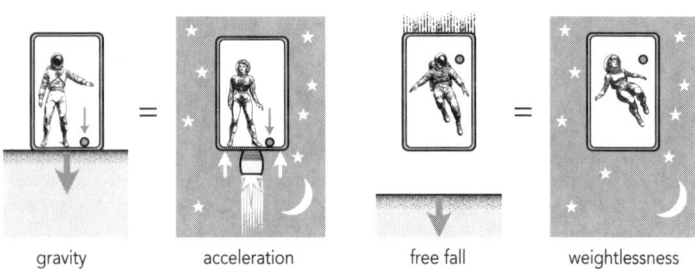

gravity = acceleration   free fall = weightlessness

ABOVE: The equivalence principle: Bob experiences gravity on Earth. Amy feels the same thing when she accelerates in a spaceship. When Amy's ship is stationary in empty space, a hammer and a feather can float weightlessly in her cabin. But as her ship speeds up, they accelerate towards the back of the ship as if in a gravitational field. To an outside observer, neither object has actually moved, only the ship.

LEFT: Astronaut Dave Scott dropping a hammer and a feather on the Moon in 1971. The experiment proved that without air resistance, both undergo equal acceleration and hit the surface at the same time despite having different masses and materials.

ACCELERATING
non-inertial frames of reference

The fact that gravity bends light would not have surprised Newton (who thought light was a stream of particles which would be affected by gravity like any other particle), but it did not seem to be consistent with the idea that light was a wave. General relativity resolved this by proposing that gravity was not a 'force' in the classical sense, but rather a curvature of spacetime caused by the presence of mass and energy.

Suppose Amy, in her spaceship moving at constant speed, shines a light beam across her cabin (*see opposite*). To Bob, on Earth, the beam appears to take a diagonal path and takes a little longer, but still hits the same point as the ship itself has moved forward.

Next Amy fires up her engines, accelerating at $10 \, m/s^2$—equivalent to the gravitational acceleration of Earth. In the time it takes for the light to cross the cabin, the accelerating ship has moved on a bit further than when it was moving at constant speed, and so the beam hits the wall a bit lower down. To Bob this is due to the ship accelerating, but Amy blames the 'artificial gravity' she experiences, as the same thing would happen if the ship was on the Earth's surface.

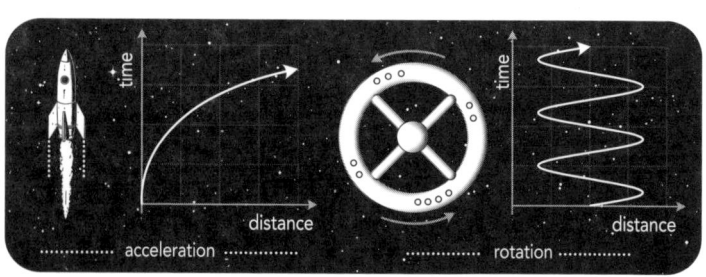

34

FACING PAGE: Spacetime diagrams of non-inertial frames of reference. In non-inertial frames, which are accelerating or rotating relative to an inertial frame, velocity changes over time, and observers within those reference frames will experience outside forces.

CURVED SPACE
unexpected angles

The paradoxes of the twins (*p.22*) or the broken rope (*p.26*) can be resolved in special relativity by changing the system of coordinates. Certainly they are real effects, but they do not fundamentally alter the assumption that space is *flat*. So what is meant by 'flat'?

Children learn that the angles of a triangle add up to 180° and that the circumference of a circle is 2π times the radius. But this is only true for surfaces with zero curvature. On the surface of a globe, the angles of a triangle add up to *more than* 180°. Suppose that explorers reached deep into the galaxy, and discovered that the sum of the surveyed angles between three stars was greater (or less) than 180°. They would have to conclude that space itself was curved (*see opposite*).

Unlike globes or saddles which are two-dimensional structures curved in a third dimension, three-dimensional space isn't curved *in* anything at all. Curved space is defined by a mathematical structure called the Riemann curvature tensor that tells us how much space is warped at any given point. In the simplest case of the space around a spherical star, the Riemann tensor can be reduced to a single number which is like the gradient of a hill. The bigger the number, the steeper the hill and the stronger the curvature.

We experience this curvature of spacetime as gravity—the greater the curvature, the stronger the gravitational effect. In fact, the picture on this page is a bit misleading; while it suggests how mass curves *space*, it can't convey the effect gravity has on *time*.

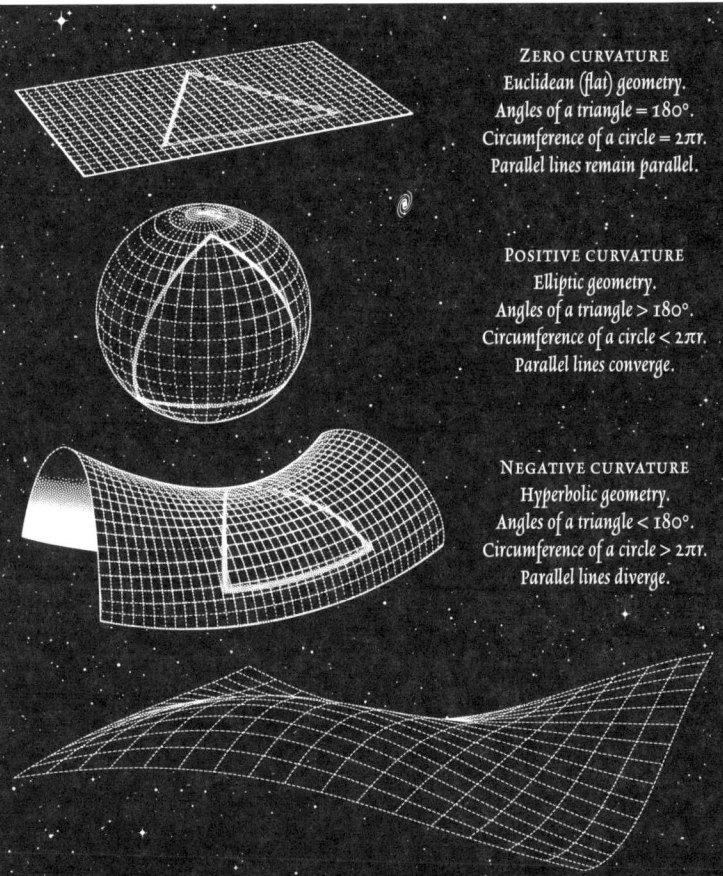

ZERO CURVATURE
Euclidean (flat) geometry.
Angles of a triangle = 180°.
Circumference of a circle = $2\pi r$.
Parallel lines remain parallel.

POSITIVE CURVATURE
Elliptic geometry.
Angles of a triangle > 180°.
Circumference of a circle < $2\pi r$.
Parallel lines converge.

NEGATIVE CURVATURE
Hyperbolic geometry.
Angles of a triangle < 180°.
Circumference of a circle > $2\pi r$.
Parallel lines diverge.

ABOVE: Non-Euclidean geometry: Only the Radius of Curvature is needed to describe a 2-dimensional surface. However, six parameters are required in the Riemann Curvature Tensor to fully describe the curvature of 3-dimensional space (the Ricci Scalar is just an approximation).
FACING PAGE: Gravity acts inwards as the space around Earth is positively curved.

Spacetime
and its curvature

When considering the effects of a massive object on the region around it, we need to regard the three space dimensions and the time dimension as equal elements in a four-dimensional structure—SPACETIME.

No less than 20 parameters are needed to describe the curvature of 4-dimensional spacetime at a point. Fortunately, the region around a spherically symmetric object like a star is much more simple, which makes it possible to separate out the spatial and temporal aspects.

In general, massive objects distort *spacetime*, not just space. When an object such as a planet or comet orbits a star, it is responding at every instant to the curvature of spacetime (*opposite top*). To paraphrase physicist John Wheeler [1911–2008]: *Matter tells spacetime how to bend; and spacetime tells matter how to move.* Einstein had solved the second part of this in 1911, when he predicted exactly how starlight would bend as it passed near to the Sun. He finally solved the first part in 1915, in his field equation of general relativity (*lower opposite*).

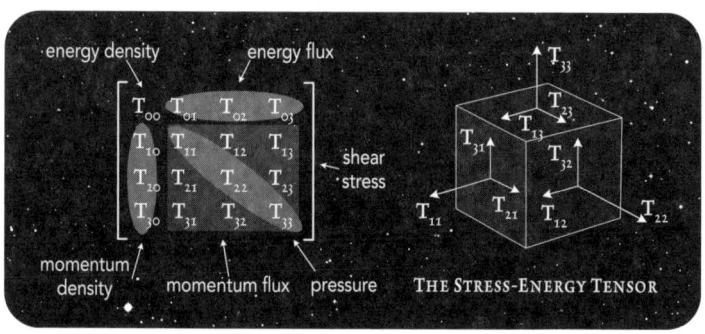

THE STRESS-ENERGY TENSOR

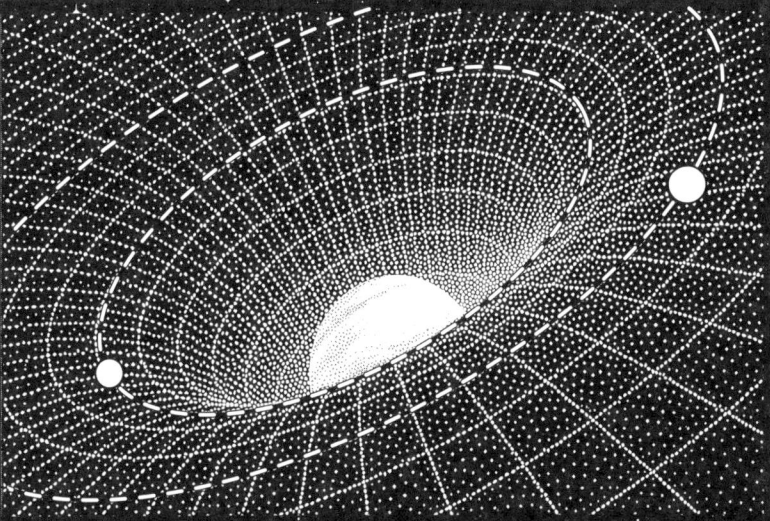

ABOVE: A massive star warps the fabric of spacetime. Planets and moons follow curved paths, falling inwards but also moving forwards, resulting in stable orbits.

$$R_{\mu\nu} - \frac{1}{2} R g_{\mu\nu} + \Lambda g_{\mu\nu} = \frac{8\pi G}{c^4} T_{\mu\nu}$$

Ricci curvature tensor · Ricci scalar curvature · metric tensor · cosmological constant · stress-energy tensor (distribution of matter and energy)

how matter-energy curves spacetime = how matter-energy moves in curved spacetime

ABOVE: Einstein's field equation, actually 10 equations in one. Tensors are mathematical objects that encode data. Λ = cosmological constant; c = speed of light; G = gravitational constant.
FACING PAGE: The stress-energy tensor $T_{\mu\nu}$ encodes the curvature of spacetime. It has 16 components, the most important being T_{00} which is determined by mass-energy density. The pressure terms T_{11}, T_{22} and T_{33} are also significant; they reveal that the gravitational influence of a neutron star under immense pressure is greater than that of a normal star of the same mass.

GRAVITATIONAL TIME DILATION
clocks in a gravity well

Imagine a circular space station which rotates at just the right speed so that astronauts in the rim experience artificial gravity equal to that of Earth. A 100m radius station would only have to rotate once every 20 seconds to achieve this. Now consider the problem of time-keeping on the station. Any clock in the rim of the space station will run more slowly than a clock placed in the central hub because of the effects of time dilation on a moving clock. Admittedly the effect will be small—but it will be there all the same. We can calculate the size of the effect as follows: if the station has a radius r and is rotating with an angular velocity of ω (the Greek letter *omega*) then the speed of the rim is $r\omega$ and the time dilation factor will be:

$$\frac{1}{\sqrt{1 - r^2 \omega^2 / c^2}}$$

Consider the situation from the point of view of an astronaut sitting in the rim. According to the principle of equivalence, they can maintain that the space station is stationary, the universe is rotating around them, and that the force pressing them into their seat is gravity. How then can they explain the fact that their clock runs more slowly than one in the hub? The only possibility is that it is gravity which causes clocks to run slow.

Mathematically speaking, they are at the bottom of a gravitational potential well whose depth ϕ (*phi*) is indicated by the energy which must be expended in lifting a mass up from the rim to the hub. For a space station of radius r spinning at an angular velocity of ω this works out to be $\phi = \frac{1}{2} v^2$ where $v = r\omega$.

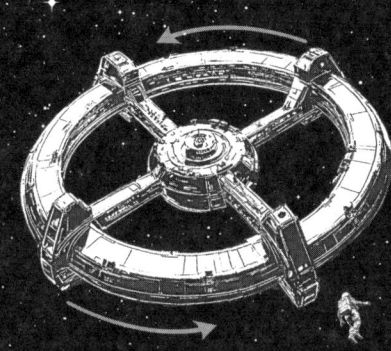

LEFT: The Ehrenfest paradox: When astronauts on a rotating space station measure its circumference, they find it is more than 2π times the radius. This is due to the relativistic contraction of their rulers out on the spinning rim. This is not standard length contraction because in their frame of reference they are not moving. However, they do feel gravity, so the resolution to this paradox is to conclude that gravity distorts space – in this case negatively.

It is not the strength of gravity which causes clocks to run slow but the depth of the gravitational potential well. The difference in gravitational potential ω is the energy needed to lift an object of unit mass from one point to another. For a rotating space station this is $\frac{1}{2}r^2\omega^2$ between the rim and the hub. In general, the time dilation factor of a potential well of depth ϕ is:

$$1/\sqrt{1 - 2\phi/c^2}$$

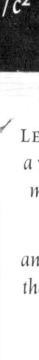

LEFT: gravitational potential at the bottom of a well of depth h on the surface of a planet of mass M and radius R is equal to gh where:

$$g = \frac{GM}{r^2}$$

and G is the gravitational constant. Clocks at the bottom therefore run slower than those at the top by a factor of:

$$1/\sqrt{1 - 2gh/c^2}$$

GEODESICS
in spacetime

In the twin paradox (*page 22*), the travelling twin returns younger than her stay-at-home brother. It turns out that when you travel from an event A to another event B, in a straight line at constant speed, you will always *maximise* the proper time taken to get there (*see page 10*). If you make any kind of deviation—by calling on another star along the way, or by travelling slowly at first then faster—then the proper time for the journey is always *less* than the straight line journey. In fact it is a general principle that, left to its own devices, any object such as a planet or a cricket ball will travel through spacetime in such a way as to *maximise* the proper time taken (*opposite top*).

We are used to principles like '*light takes the path through a lens which minimizes its time of travel*' and '*airplanes fly great circle routes because they are the shortest paths from one city to another*'. It turns out that all these principles are essentially the same, and that it does not matter whether the chosen quantity is maximized or minimized—all these paths are GEODESICS.

Any reference frame which follows a geodesic path through spacetime becomes what is called a 'locally inertial frame'. Objects within this frame behave as if they were in empty space, no matter what the curvature/gravitation.

This is why astronauts in orbit experience weightlessness as if they were floating freely in space, even though they are still within a gravitational field (*right*).

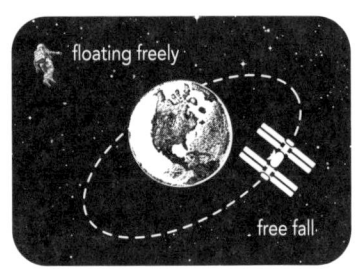

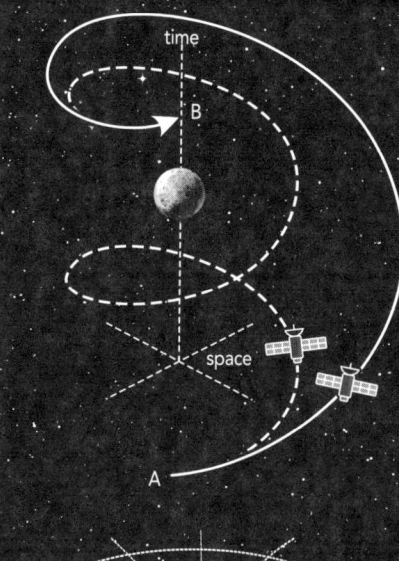

LEFT: The trajectories of two satellites orbiting a planet. Both are launched at A and reach B at the same time. The first makes two small circular orbits, the other a large elliptical one. Each orbit maximizes proper time since any deviation results in a smaller proper time. Both are therefore valid geodesics, even though the proper time for the elliptical orbit is larger than that of the circular orbit. Although the paths look curved in 3D space, they are in fact 'straight' lines through 4D spacetime.

ABOVE: Light ray geodesics around a black hole.
OPPOSITE: Satellites are in constant free fall, their speed maintaining orbit by tracing a geodesic path. Inside, astronauts feel weightless as geodesic motion masks the gravitational pull.

ABOVE: A thrown ball decelerates on the way up and accelerates on the way down as it follows a parabolic geodesic through spacetime which maximizes its proper time. Fast balls just follow a section of the same path as slow balls.

GRAVITATIONAL LENSING
the bending of starlight

General relativity's prediction that light follows geodesic paths through a curved spacetime was spectacularly proven during the solar eclipse of 1919, when Sir Arthur Eddington [1882–1944] and his team observed the light from a known star bending from its straight path as it came under the influence of the Sun's gravity (*opposite top*).

GRAVITATIONAL LENSING occurs when a large mass—like a galaxy or galaxy cluster—bends the light travelling from a distant light source (e.g. a quasar or galaxy) towards the observer. Light bends for two reasons—firstly, the rays which pass near to the mass travel through regions of low gravitational potential where time passes more slowly, in effect lagging behind those that pass by at a greater distance. Secondly, owing to the warping of space caused by the non-uniform gravitational field, these rays also have further to travel (*lower opposite*).

Lensing can take several forms: STRONG LENSING creates the dramatic arcs and multiple images of *Einstein rings* (*below*).

WEAK LENSING is a more subtle background distortion, magnifying faint, distant objects or revealing large-scale gravitational effects.

MICROLENSING appears as a temporary brightening when a compact mass passes in front of the source, aiding in the discovery of exoplanets, brown dwarfs and other difficult to detect celestial inhabitants.

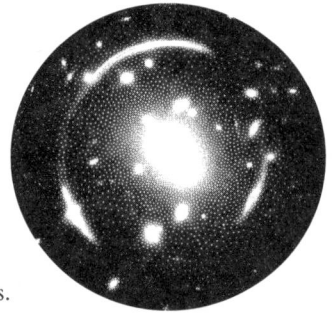

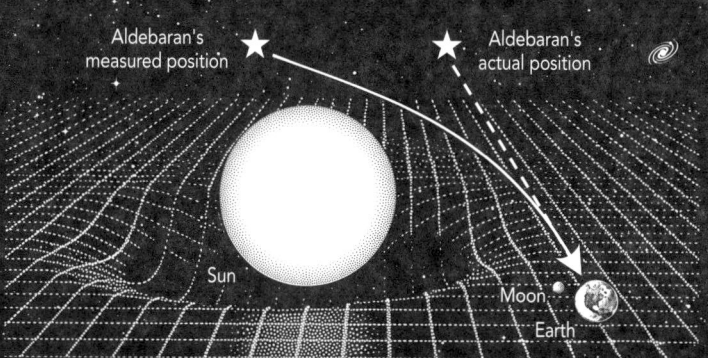

ABOVE: *Eddington's experiment measuring the deflection of starlight during the 1919 solar eclipse confirmed Einstein's prediction that the apparent position of the stars would shift away from the Sun by 1.75 arcseconds – equivalent to the width of a penny a mile away.*

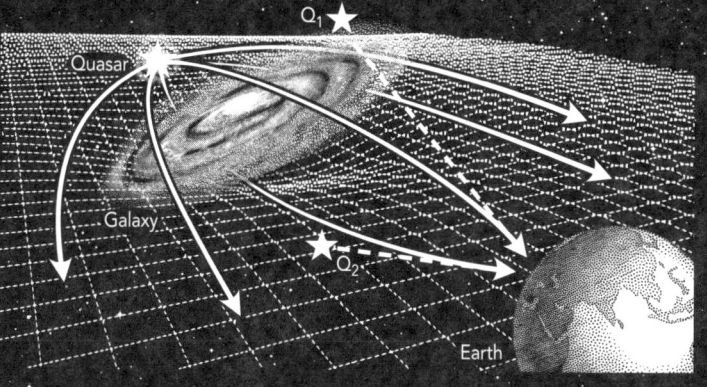

ABOVE: *A gravitational lens: a distant quasar is lensed by a foreground galaxy to produce multiple images (Q_1, Q_2, etc). The time delays between the light travel paths of different images can be used as a measure to calculate universal expansion.*

RELATIVISTIC ORBITS
the precession of Mercury

For centuries, astronomers had puzzled over the orbit of Mercury. According to Newton's laws it should stay in a constant elliptical orbit, but instead the ellipse rotates (or *precesses*), like the unfolding petals of a flower (*opposite top*). Some of the precession could be blamed on the perturbing influences of other planets, but not all.

In 1915, Einstein solved the problem. His new theory of general relativity predicted that massive objects would warp the fabric of spacetime around it. This curvature is not just a three-dimensional effect (like a ball on a rubber sheet) but extends to the four dimensions of space and time. The spacetime geodesics around a massive object are thus not the perfect ellipses of Newtonian mechanics, but are more complex curves that change over distance (*lower opposite*), and the greater the mass, the more pronounced the curvature.

When it comes to supermassive objects, things get really strange. In the case of black holes, there are no stable orbits within 3 times the Schwarzschild radius. Even outside this forbidden zone, there are some very odd orbits which come in, circle several times round the black hole, and then go out again (*see examples below*)!

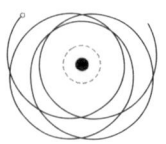
modest precession

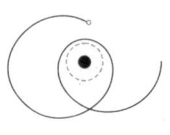
twice orbital period

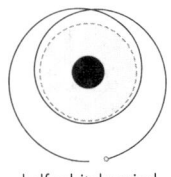
half orbital period

inspiral

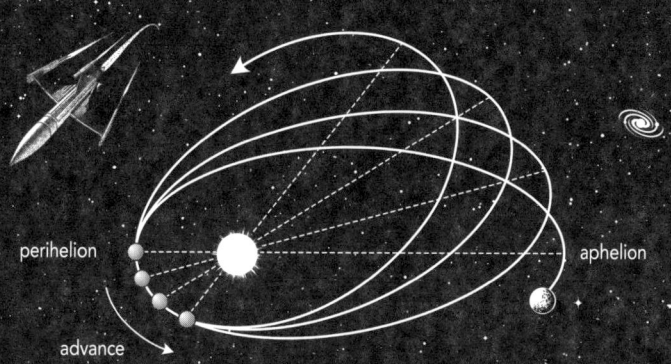

ABOVE: The precession of Mercury: The Sun's mass warps spacetime causing deviations in the orbits of nearby planets. The curvature affects both space (a planet's position) and time (its speed), resulting in the planet 'rolling' along the fabric of spacetime. In the solar system, Mercury is seen to have the greatest orbital precession due to its proximity to the Sun.

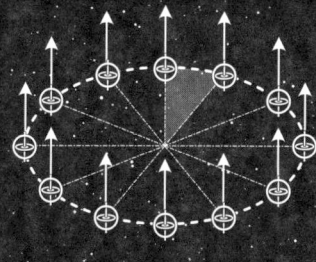

LEFT: A gyroscope spinning in empty space will always point in the same direction, even if it follows a circular path.

RIGHT: A massive body warps spacetime. An orbiting gyroscope traces a conical path, causing its axis of rotation to slowly precess.

TIME DILATION NEAR A STAR
the nearer you get, the slower you go

The amount of TIME DILATION on the surface of a planet is directly related to the planet's *escape velocity* (the speed at which a projectile must travel if it is to escape from the planet's gravity). That is why clocks on the Moon do not go quite as slow as clocks on Earth (*below*)—an important factor when it comes to implementing a lunar GPS system.

The escape velocity v_{esc} of a planet of radius R and mass M is:

$$\sqrt{\frac{2GM}{R}}$$

where G is the Newtonian constant of gravitation and $\frac{GM}{R}$ is the gravitational potential ϕ (*phi*). This means that, using the time dilation formula (*page 41*), the time dilation factor at the surface of a planet is:

$$\frac{1}{\sqrt{1 - 2GM/Rc^2}} = \frac{1}{\sqrt{1 - v_{esc}^2/c^2}}$$

In the case of an object so dense and massive that its escape velocity equals the speed of light, time dilation becomes infinite and, from the perspective of a distant observer, time effectively stops. Such an object would, in fact, be a black hole.

Mercury 8.6 μs/day · Earth 60.2 μs/day · Moon 2.7 μs/day · Jupiter 1705 μs/day · Saturn 605.7 μs/day · Neptune 265.6 μs/day · Venus 51.6 μs/day · Mars 12.1 μs/day · Uranus 218.2 μs/day

TIME DILATION

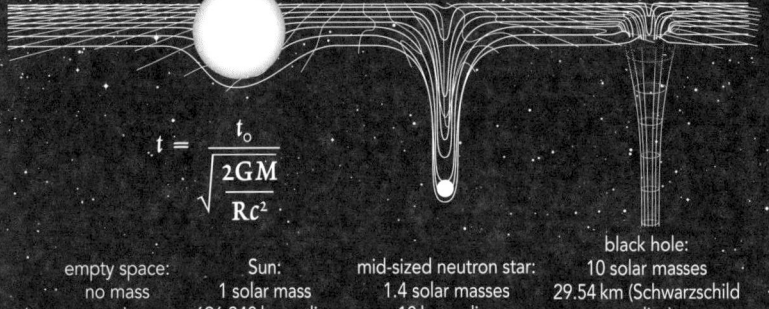

$$t = \frac{t_o}{\sqrt{\dfrac{2GM}{Rc^2}}}$$

empty space:	Sun:	mid-sized neutron star:	black hole:
no mass	1 solar mass	1.4 solar masses	10 solar masses
no size	696,340 km radius	10 km radius	29.54 km (Schwarzschild radius)
0 secs	183.4 ms/day	1.3 days/day	infinite

ABOVE & FACING: *Gravitational time dilation for different celestial objects. The amount that time slows is shown per day as compared to empty space which is free from gravitational effects.*

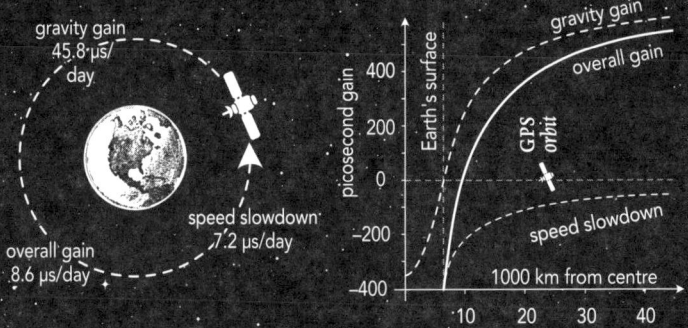

ABOVE: *Earth orbit time dilation: GPS satellites are fast-moving (time slows down), but also orbit high above the Earth and experience less gravity than the surface (time speeds up). To keep their clocks accurate, they have to compensate for the effects of both special and general relativity.*

THE SCHWARZSCHILD RADIUS
over the horizon

Provided you are some distance from a star, the effects of the curvature of space can be approximated by Newton's law of gravity. But imagine a system with a particularly massive central star. Bob remains in orbit while Amy takes a closer look. The nearer she gets to the star, the greater the gravitational potential difference ϕ between her and Bob, and the slower her clocks go compared to his (*see page 41*).

The formula for gravitational time dilation $1/\sqrt{(1 - 2\phi/c^2)}$ makes a curious prediction. When ϕ increases to where $2\phi/c^2 = 1$, the denominator becomes zero and, according to Bob, Amy's clocks will apparently stop completely! The point at which this happens is the *event horizon*. Derived by physicist Karl Schwarzschild [1873–1916] from his first exact solution to Einstein's field equations, the **SCHWARZSCHILD RADIUS** R_S of the event horizon for a star of mass M is:

$$R_S = \frac{2GM}{c^2}$$

(where G is the Newtonian constant of gravitation).

If Amy was foolish enough to venture this far, she would not notice her clocks running slow, nor would she necessarily detect anything unusual—the event horizon is just a consequence of an outside point of view. She would, however, be committing to a one-way trip because any star massive enough and small enough to fit inside its own event horizon has already become a **BLACK HOLE** (*right*).

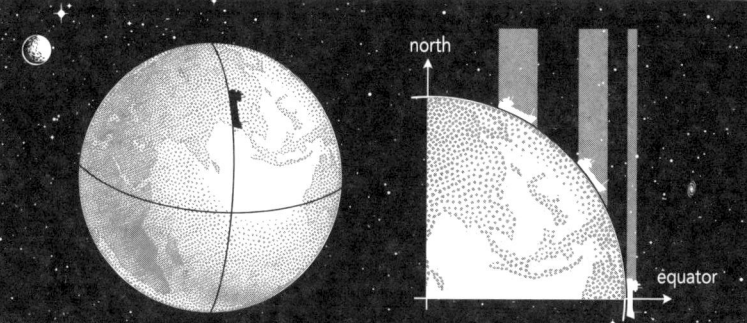

ABOVE: A ship travels due south towards the equator. An observer high above the North Pole perceives the ship getting shorter and shorter until it finally vanishes, but the sailors on board experience nothing unusual happening at all. The equator is a 'coordinate singularity' where the shortening becomes infinite—but only from the polar observer's point of view.

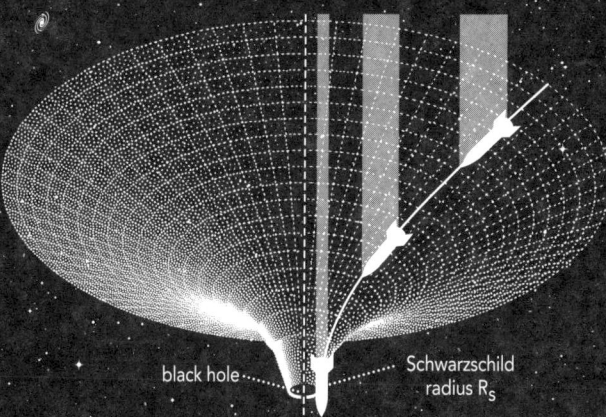

ABOVE: Similarly, from Bob's point of view, spacetime near a black hole follows the curve of a Flamm's paraboloid—a funnel-shaped parabola which becomes vertical when $r = R_S$. To Bob, Amy's probe gets increasingly shorter and slower as she nears the event horizon (which she reaches in a finite time). Bob never sees Amy pass over it, she just gradually fades from sight.

BLACK HOLES
the trick of singularity

One of the more extreme solutions to Einstein's field equations describes a mass collapsing to a single infinitely dense point—a spacetime SINGULARITY, or black hole, where known physical laws cease to function. A stellar black hole is formed when a star of at least 8 times the Sun's mass runs out of fuel, causing it to implode from overwhelming gravitational pressure. This cannot be observed, as before it actually happens, the star cloaks itself inside its event horizon.

Around a black hole, the fabric of spacetime undergoes extreme distortions leading to gravitational time dilation, with time slowing dramatically near to the event horizon. The intense pull causes GRAVITATIONAL REDSHIFT, stretching light to longer, redder wavelengths (*below*). Light from just outside the horizon may eventually escape (albeit massively redshifted) but all light within is trapped forever. The environment around a black hole is fearsomely turbulent, with powerful X-ray emissions and relativistic plasma jets fired at near light speed from the hole's poles, driven by strong magnetic fields and the infall of matter from the surrounding accretion disk (*opposite*).

Black holes are sometimes described as cosmic vacuum cleaners, sucking in everything around them. However, if you came across one in space you would have no more to fear than an encounter with a star of the same mass, and could orbit it from a distance in perfect safety.

$$\lambda_{observed} = \lambda_{original} \frac{1}{\sqrt{1 - R_s/r}}$$

increasing wavelength

ACCRETION DISC: intense gravity draws in a swirling vortex of gas and debris, heating the material to extreme temperatures and causing X-ray and radiation emissions.

RELATIVISTIC JETS: plasma streams are launched at near light speeds from the poles by twisting magnetic fields.

PHOTON SPHERE: a bright ring of trapped photons orbiting at relativistic speeds just outside the event horizon.

INNERMOST STABLE ORBIT: the closest matter can orbit the black hole, inside which any matter spirals inward.

The Schwarzschild radius (R_S) of a black hole with the same mass as the Earth is 8.9 mm, roughly the size of a pea. The R_S of the Sun is 2.9 km, similar in size to a small mountain. The R_S for a supermassive black hole of 4.3 million solar masses, like Sagittarius A* at the centre of our galaxy, is 12.7 million km – about one fifth the radius of Mercury's orbit.

OPPOSITE: Gravitational redshift: Light escaping a black hole loses energy. However, because the speed of light is constant, any energy loss is manifested as an increase in its wavelength λ, making it appear redder from a distance; r is black hole radius, R_S is the Schwarzschild radius.

INTO A BLACK HOLE
beyond the horizon

Although it might be possible to pass through the event horizon of a black hole, there are several factors to consider before setting out. The first is, obviously, we will not return. Second, choose a really big black hole, as a point source of gravity would attract our feet more than our ears and the resulting differential (or tidal) force would spaghettify you into long strands. Third, find a black hole in a suitably empty region of space, otherwise the intense stream of high-energy particles of gas and dust streaming into the hole will fry us.

The first thing to notice as we begin our descent is that, as light bends significantly near the hole, we see a warped image of the stars around and behind the hole. As we descend further, the hole's 'surface' no longer appears below us but curves upwards at the edges, pushing the stars into a smaller and smaller disc. By now, the stars are shining in the hard ultraviolet and X-ray regions of the spectrum because of the intense gravitational blueshift.

To observers back at home, our redshifted clocks appear to slow to a stop; it follows that for us, time outside is going ever faster and so, even if we could escape, we would return to an infinitely old universe. So old, in fact, that the very black hole in which we are trapped might have evaporated and ceased to exist.

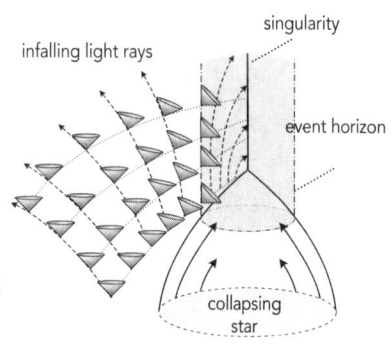

As we cross the event horizon, the circle of light from the stars and galaxies we left behind plays through the history of the universe in a flash of gamma rays and is gone. Everything inside the ship behaves normally – the rockets seem to hold us just below the horizon against the modest gravity. But now it would take an infinite amount of energy to arrest our fall and so even hovering is not an option. The rules of spacetime have changed, and motion in any direction speeds our journey to the singularity. All we can do is switch off the engines and enjoy the ride until the tidal forces tear us apart.

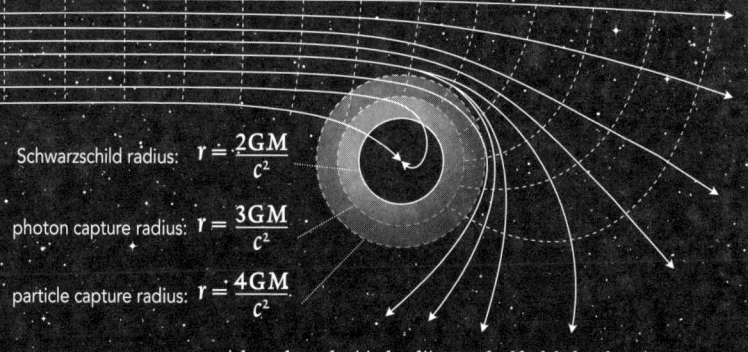

Schwarzschild radius: $r = \dfrac{2GM}{c^2}$

photon capture radius: $r = \dfrac{3GM}{c^2}$

particle capture radius: $r = \dfrac{4GM}{c^2}$

ABOVE & OPPOSITE: Light paths and critical radii around a black hole. There is no escaping the fierce grasp of gravity – even light becomes tightly wrapped in a photon sphere.

Special Effects
drags, waves & shifts

General relativity's dynamic relationship between mass-energy and spacetime curvature predicts that the rotation of a large body gives an additional twist to spacetime. This FRAME-DRAGGING pulls nearby objects and light paths along in the direction of rotation. Although usually small, the effects become much greater around massive rapidly rotating objects like neutron stars or black holes (*opposite top*).

Gravitational influence propagates through space at the speed of light in the form of GRAVITATIONAL WAVES, generated by the most energetic processes in the universe: the merger of massive objects like neutron stars and black holes, or the vast explosions of supernovae. Predicted by Einstein in 1916, their existence was confirmed in 2015 when waves from a black hole merger were detected (*opposite centre*).

When light waves travel away from an observer, wavelengths stretch toward the red end of the spectrum (*lower opposite*). Galaxies exhibit increasing redshifts with distance, but this COSMOLOGICAL REDSHIFT stems from space's uniform expansion, rather than motion. GRAVITATIONAL REDSHIFT (*below*) occurs when light escapes a strong gravitational field, stretching to longer wavelengths, while light falling into a gravitational field is blueshifted to shorter wavelengths.

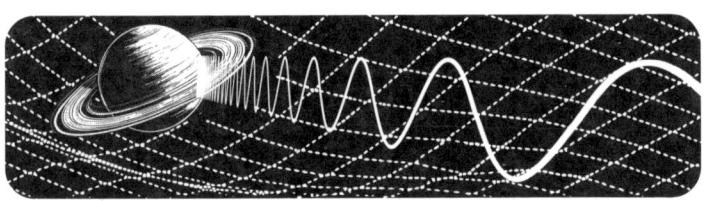

ABOVE: Frame dragging (the Lense-Thirring effect) occurs when the rotation of a massive object distorts the spacetime around it, causing spacetime itself to swirl in a manner akin to how a rotating object in a nice hot cup of tea draws the surrounding fluid with it.

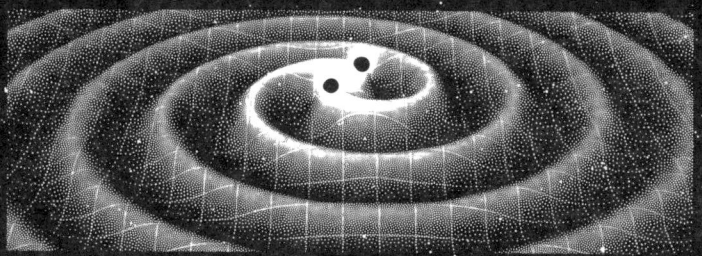

ABOVE: Merging black holes, warping the curvature of spacetime. The asymmetric accelerations of massive objects cause rippling gravitational pressure waves propagating outward at light speed.

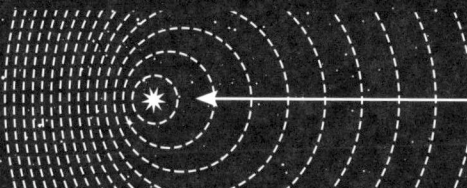

wavelength of light emitted from a star receding at velocity v is increased by a factor of:

$$\sqrt{\frac{c+v}{c-v}}$$

ABOVE: Relativistic Doppler Shift is a consequence of special relativity, and is due to the relative motion between a light source and an observer. It occurs when stars, galaxies, or other celestial objects are moving toward (blueshift) or away from (redshift) the observer.

RELATIVISTIC FORMULÆ

SPEED OF LIGHT: c as derived from Maxwell's equations where:
ε_0 is the vacuum permittivity $= 8.854 \times 10^{-12}$ Fm^{-1}
μ_0 is the vacuum permeability $= 1.256 \times 10^{-6}$ Hm^{-1}

$$c = \frac{1}{\sqrt{\varepsilon_0 \mu_0}} = 299{,}792{,}458 \text{ ms}^{-1}$$

GRAVITATIONAL CONSTANT:

$G = 6.674 \times 10^{-11}$ m^3 kg^{-1} s^{-2}

LORENTZ FACTOR: $\gamma = 1/\sqrt{1 - v^2/c^2}$

SCHWARZSCHILD RADIUS (R_s) of a black hole of mass M is $2GM/c^2$ where G is the gravitational constant.

SPACETIME INTERVAL: the separation between two events in spacetime, invariant for all observers:
$$s^2 = -c^2(\Delta t)^2 + (\Delta x)^2 + (\Delta y)^2 + (\Delta z)^2$$

PROPER TIME: the time measured by a clock moving with an object in its own rest frame. The interval between two events in any frame is:
$\sqrt{(\Delta t^2 - \Delta x^2/c^2 - \Delta y^2/c^2 - \Delta z^2/c^2)}$.

GRAVITATIONAL TIME DILATION: A clock at height h runs faster than a clock on the ground by a factor equal to $1/\sqrt{1 - 2gh/c^2}$ where g is the local acceleration due to gravity. In general, if the gravitational potential difference between two points is equal to $\Delta\phi$ then the time dilation factor will be: $1/\sqrt{1 - 2\Delta\phi/c^2}$.

RELATIVISTIC MASS, MOMENTUM & KINETIC ENERGY: An object with rest mass m_o moving at velocity v will have mass γm_o; its momentum will be $\gamma m_o v$ and its kinetic energy will be $m_o c^2(\gamma - 1)$. Its total relativistic energy is $m_o c^2$.

RELATIVISTIC ABERRATION: Light emitted by a moving object is concentrated conically towards its direction of motion by an angle θ where:
$\cos\theta_o = \cos\theta_s - \frac{v}{c} / 1 - \frac{v}{c} \cos\theta_s$

GRAVITATIONAL REDSHIFT: The wavelength of light λ_o from an object of radius r is $\lambda_e/\sqrt{(1 - R_s/r)}$ where R_s is the Schwarzschild radius r and λ_e is the emitted wavelength.

In the following equations, frame A is assumed to be moving along the x-axis at velocity v with respect to 'stationary' frame B.

LORENTZ TRANSFORMATION:

An event with coordinates (t, x, y, z) in B coordinates $(\gamma(t - vx/c^2), \gamma(x - vt), y, z)$ in A.

LENGTH CONTRACTION: A ruler length l oriented along the x-axis in A has length l/γ in B.

TIME DILATION: A clock which records a time t in A will record a time γ t in B.

TIME ON A DISTANT STAR: Moving towards a star at a distance x at a speed v, time t on that star will move forward by $t = xv/c^2$.

ADDITION OF VELOCITIES: An object which has a speed u in A will have a velocity $(u + v)/(1 + uv/c^2)$ in B.

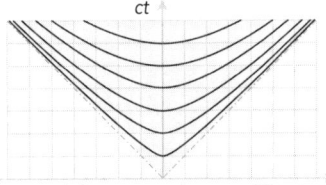

Hyperbolic contours mark events equidistant in proper time from the origin.